Studies in Computational Intelligence

Volume 684

Series editor

Janusz Kacprzyk, Polish Academy of Sciences, Warsaw, Poland
e-mail: kacprzyk@ibspan.waw.pl

About this Series

The series "Studies in Computational Intelligence" (SCI) publishes new developments and advances in the various areas of computational intelligence—quickly and with a high quality. The intent is to cover the theory, applications, and design methods of computational intelligence, as embedded in the fields of engineering, computer science, physics and life sciences, as well as the methodologies behind them. The series contains monographs, lecture notes and edited volumes in computational intelligence spanning the areas of neural networks, connectionist systems, genetic algorithms, evolutionary computation, artificial intelligence, cellular automata, self-organizing systems, soft computing, fuzzy systems, and hybrid intelligent systems. Of particular value to both the contributors and the readership are the short publication timeframe and the worldwide distribution, which enable both wide and rapid dissemination of research output.

More information about this series at http://www.springer.com/series/7092

Krzysztof Cpałka

Design of Interpretable Fuzzy Systems

Krzysztof Cpałka
Institute of Computational Intelligence
Częstochowa University of Technology
Częstochowa
Poland

ISSN 1860-949X ISSN 1860-9503 (electronic)
Studies in Computational Intelligence
ISBN 978-3-319-85006-1 ISBN 978-3-319-52881-6 (eBook)
DOI 10.1007/978-3-319-52881-6

Printed on acid-free paper

This Springer imprint is published by Springer Nature
The registered company is Springer International Publishing AG
The registered company address is: Gewerbestrasse 11, 6330 Cham, Switzerland

Preface

Fuzzy systems are well suited for advanced scientific and engineering applications. They are universal approximators and the knowledge stored in them is represented in the form of fuzzy rules. Initially, such rules were thought to be inherently interpretable. That resulted, among others, from the theory of fuzzy sets proposed by Prof. Lotfi Zadeh, which assumed that fuzzy rules used fuzzy sets and interpretation of a fuzzy set is intuitive. However, in the 1990s researchers drew attention to the fact that the use of a large number of fuzzy rules and sets increases accuracy of system operation but significantly reduces readability of a rule-based notation. Therefore, solutions associated with the term "interpretability" began to appear. Currently, this term goes far beyond the notions of the fuzzy set and the fuzzy rule. The purpose of this book is to show different aspects of interpretable fuzzy systems design in various fields of applications. This book addresses issues of fuzzy sets and systems theory, machine learning, evolutionary algorithms inspired by nature, multi-criterion optimization, classification, biometrics, automatics and control theory.

The book is the result of collaboration with colleagues from the Institute of Computational Intelligence at Częstochowa University of Technology. The author would like to thank Dr. Łukasz Bartczuk, Dr. Krystian Łapa, Dr. Andrzej Przybył and Dr. Marcin Zalasiński for their cooperation. The author specially thanks Prof. Leszek Rutkowski for his invaluable help and great kindness, which contributed to the writing of this book. The author also thanks his wife Agnieszka, children, Malwina and Patryk, and his parents, Wanda and Józef, for inspirational motivation, heartfelt support and patience.

Częstochowa, Poland
October 2016

Krzysztof Cpałka

Acknowledgements

This book was partly supported by the National Science Centre (Poland) on the basis of the decision number DEC-2012/05/B/ST7/02138.

Contents

Chapter 1
Introduction

Systems using the fuzzy set theory [127] are excellent for advanced research and engineering applications in the field of classification, nonlinear modelling, identification, control, prediction, medical diagnostics, industrial diagnostics, economics, biometrics, image processing, etc. (see e.g. [1–8, 10–13, 15–42, 44, 46–50, 52–61, 63, 64, 67–74, 76–95, 97, 99, 100, 102–107, 110, 112–120, 122, 123, 125, 126, 128, 130, 131]). Such systems, henceforth referred to as fuzzy systems, have several important advantages that distinguish them from other alternative solutions. Their advantages can be summarized as follows:

- Fuzzy systems allow us to describe phenomena and concepts that are ambiguous and imprecise. In practice, it is such phenomena and concepts that occur most often, so using the classical sets theory and bivalent logic is not a proper solution in these cases.
- Fuzzy systems use a very intuitive rule notation in the form of $if \ldots then \ldots$, which consists of antecedents and consequences of the rules, and input and output linguistic variables among others. Particularly important are linguistic systems (see e.g. [65, 66, 96, 97]), in which the values of linguistic variables are represented by fuzzy sets. Knowledge accumulated in these systems provides very good opportunities for its interpretation.
- Fuzzy systems significantly support nonlinear modelling. In particular, they allow us to extract readable knowledge (in the form of fuzzy rules) from phenomena and objects for which a model is still not known or it is a so-called black box (see e.g. [43, 109]). Motivation to develop an interpretable and accurate model is driven by the desire to provide predictability, which guarantees widely understood safety, reduces operating costs, provides an ability to control, explains the principle of operation, etc. It is particularly important that models which are accurate and easy to interpret, and which result from modeling based on the fuzzy set theory can actually be created.
- Fuzzy systems can be combined with other approaches, thus creating various kinds of hierarchical systems or systems encoding fuzzy rules in the form of tree structures (see e.g. [101]). Such systems are one of the basic tools for data mining.

K. Cpałka, *Design of Interpretable Fuzzy Systems*, Studies in Computational Intelligence 684, DOI 10.1007/978-3-319-52881-6_1

The purpose of data mining comprises, among others, an automatic discovery of unknown and informal dependencies, tendencies and anomalies in data sets and their readable presentation by, for example, using clear fuzzy rules. This sphere of application results in new and increasingly sophisticated requirements for classification methods. They need to, among others, be able to work in larger ensembles and properly cooperate in those teams. An advantage of using classifier ensembles also includes a possibility of achieving better accuracy of operation and a more precise decision mechanism, which makes design problems less important.

- Fuzzy systems provide a good framework for creating a variety of extensions which, for example, enable processing of imprecise information (uncertainty). Such extensions are used in systems based on the terms of a so-called rough-fuzzy set, fuzzy-rough set, interval-valued fuzzy set, fuzzy-valued fuzzy set, etc. (see e.g. [70, 81, 82, 108]). Uncertainty processed by those systems generally results from limited perception, differences in decisions made by expert teams, the specificity of available input data, which may be incomplete or affected by a discretization error, etc. Extensions of fuzzy systems aimed at processing of uncertainty are extremely useful tools in certain areas of application but they have one common characteristic. They usually add extensive information on reliability of a generated response to the system output, which makes it harder to take care of readability of the knowledge stored in them.
- Fuzzy systems may (but do not have to) use knowledge and skills of so-called experts. Such persons can customize a formulated description of a considered problem to the specifics of fuzzy rule notations. It could be the basis for an appropriate initialization of a fuzzy rules base, which creates unique abilities in the context of other modelling methods.
- Fuzzy systems are able to extract knowledge from numeric data. In contrast to the initialization of a fuzzy rules base involving an expert, in practice, another method of a fuzzy rule base creation is used much more frequently. It is based on a so-called learning sequence, which consists of learning sets. Each of these sets consists of a set of input signals and their corresponding output reference signals (so-called teacher signals). A medical record is a good example of such learning sequence. In such documentation learning sets are represented by encoded numeric medical data of individual patients and corresponding medical diagnoses (made by an expert). A learning sequence should contain a sufficient number of learning sets (i.e. it should be representative) so as to fully cover the problem space. Referring to the above mentioned medical system, as a result of a learning process using a sequence created on the basis of a medical record, the system has a chance to help a doctor in making a diagnosis. Such system applications prove to work well in practice. However, due attention should be drawn to some important aspects. Firstly, the purpose of fuzzy systems design is not an irrational attempt to diminish the role of experts - fuzzy systems usually perform the role of support systems. Secondly, extraction of knowledge during a learning process has important cognitive benefits. Thirdly, learning sequences usually come from a background process of non-invasive identification conducted on-line, which has nothing to do with the above mentioned generally well-understood medical analogy.

- Fuzzy systems can be implemented in hardware. Support in this area could include the use of FPGA chips (Field Programmable Gate Array). HIL (Hardware-In-the-Loop) emulators created with the use of these chips allow us, among others, to perform cheap and safe tests of complex control systems, for which experimenting on a real object is not advisable. Since the first hardware solutions supporting fuzzy systems were introduced [124], other technologies have also been used in this field, e.g., CMOS (Complementary Metal-Oxide Semiconductor), VLSI (Very Large Scale Integration), etc. (see e.g. [129]).
- Fuzzy systems can be easily adapted to create an online fuzzy rules base. It is particularly important in the case of issues of nonlinear modelling performed in cooperation with non-invasive online identification (for the object). In this case an efficient RTE interface is very important (Real-Time Ethernet Class 1, 2 or -the most efficient- Class 3), e.g., ECAT, Sercos III, Varan Bus, Ethernet Powerlink, Profinet IRT, etc. Such interface makes it possible to create a fuzzy rules base in a sufficiently long time horizon in order to increase modeling precision.
- Fuzzy systems support operation and analysis of Internet social networks. Such networks have recently gained in importance and have become an indispensable form of communication. Prediction of missing connections, activity analysis and modelling aspects are only some of the examples of application areas in which fuzzy systems are an important tool (see e.g. [35, 93]).

It is worth noting that in the literature those fuzzy systems which can be learned are called neuro-fuzzy systems. This is due to the use of the algorithms used for learning artificial neural networks in the learning process of these systems. However, it should be emphasized that metaheuristic evolutionary algorithms are also being currently used for fuzzy systems learning.

In this book we consider practical aspects of designing fuzzy systems aimed at broadly defined interpretability. The purpose of this book is to show that the term "interpretability" goes far beyond the concept of readability of a fuzzy set and fuzzy rules. The individual chapters describe various aspects of interpretability, placing them in specific and varied application areas.

Chapter 2 describes selected issues in the field of fuzzy systems. Because Professor Zadeh's theory of fuzzy sets and systems [127] has been described in numerous scientific and popular science publications, in Chap. 2 only those issues which are essential to proper understanding of the subsequent chapters are presented. We place a particular focus on so-called Mamdani-type fuzzy systems and logical-type fuzzy systems, which are described later in this book (see e.g. [96–98]). However, it should be noted that the solutions presented in this book are universal and can be successfully applied to other varieties of fuzzy systems which are well known in the literature (e.g. ANFIS [45], ANNBFIS [21], DENFIS [51], FALCON [62], GARIC [9], NEFCLASS [74], NEFPROX [74, 75], SANFIS [121], etc.). In the final part of the chapter its bibliography is presented.

In Chap. 3 an introduction to the interpretability of fuzzy systems is presented. It places a special focus on the problem, frequently reviewed in the literature, of a compromise between system accuracy and interpretability of the knowledge accumulated in such system, and the relationship between complexity of a system and semantic interpretability of its knowledge. Moreover, Chap. 3 offers an overview of typical solutions in the field of interpretability of fuzzy systems and a review of the basic criteria (measures) of fuzzy systems interpretability which are under discussion in the literature. These criteria can be used for two purposes. First of all, they can be used to evaluate readability of fuzzy systems which, e.g., have already been learnt. However, using them in the process of fuzzy systems design seems a much more interesting application. Such use of the criteria allows us to comprehensively affect a system structure and its rules, so the criteria do not perform only an informative function. In the final part of the chapter its summary and bibliography are presented.

Chapter 4 describes how appropriate selection of the structure of a fuzzy system might affect its interpretability. It specifically focuses on precise operators of aggregation, inference, and defuzzification. These operators are used in relation to Mamdani-type systems and logical-type systems presented in Chap. 2. As a result of this relation, so-called flexible neuro-fuzzy systems are obtained (see e.g. [97]). From the point of view of the issues discussed in this book they have a very important feature: by using precision operators which allow us to move from the notation of fuzzy rules to the network structure of a fuzzy system, we can achieve the required accuracy using a less complex rule base. The correctness of this approach seems obvious, but so far it has not been considered in the literature in this particular context. Moreover, in Chap. 3 a review of expanded interpretability criteria, tailored to the specific type of flexible systems, is presented. These criteria complement the measures described in Chap. 2. In the final part of the chapter simulation results along with its summary and bibliography are presented.

Chapter 5 describes the issue of influencing fuzzy systems interpretability when these systems are learned using gradient algorithms. Its special focus is, among others, on the appropriate initialization and learning procedures of a fuzzy system, which may affect the accuracy of its operation. This results from the fact that an unsatisfactory system accuracy tends to expand its fuzzy rules base, which reduces its readability. Moreover, Chap. 5 describes interesting algorithms used to improve interpretability, which may be executed after the learning process has finished. Their purpose is to eliminate those system components (such as inputs, rules, fuzzy sets, etc.) whose reduction does not adversely affect system accuracy. Interestingly, reduction can also positively affect the obtained accuracy. It happens when system components causing contradictions in a rule base are eliminated. The group of the methods under consideration comprises the ones aimed at maximizing the reduction of the system level and at maximizing the final accuracy of a system. It should be emphasized that the reduction of a system also means merging of similar (according to the assumed similarity measure) fuzzy sets. In the final part of the chapter simulation results, its summary and bibliography are presented.

Chapter 6 discusses the issue of affecting fuzzy systems interpretability when these systems are learned using evolutionary algorithms. It specifically focuses, among others, on presenting the principles of construction of hybrid, metaheuristic population algorithms. Hybridity of these methods creates excellent opportunities for a comprehensive design of fuzzy systems. It allows us, among others, to automatically choose the number of rules, select a subset of the most important input attributes, affect the behavior of inference and aggregation operators, etc. In Chap. 6 we describe two different types of hybrid algorithms, referring to the two most common groups of evolutionary algorithms, i.e., one-population algorithms and multi-population ones. In the final part of the chapter exemplary simulation results along with the chapter's summary and bibliography are presented.

Chapter 7 describes interpretability of fuzzy systems applied to nonlinear modelling and control. In particular the problem of nonlinear dynamic objects modelling and the problem of control of a CNC machine with a jerk limit are described. In the final part of the chapter exemplary simulation results and the chapter's summary along with its bibliography are presented.

Chapter 8 deals with the issue of influence on fuzzy systems interpretability by elimination of supervised learning. We present two sample algorithms of fuzzy system initialization on the basis of input data with no need for supervised learning. They may be used, for example, to verify one's identity on the basis of the characteristic features of their handwritten signature and characteristic signature areas. However, they can be used in all areas of application in which a comparison of the signals changing over time and having a specific interpretation is performed. In the final part of the chapter exemplary simulation results and the chapter's summary together with its bibliography are presented.

The last chapter of the book is Chap. 9, which contains concluding remarks and future perspectives. The book ends with an index of the most important phrases used in the book.

References

1. Abonyi, J.: Fuzzy Model Identification for Control. Birkhäuser, Switzerland (2003)
2. Adeli, H., Hung, S.L.: Machine Learning: Neural Networks, Genetic Algorithms, and Fuzzy Systems. Wiley, New Jersey (1994)
3. Babuska, R.: Fuzzy Modeling for Control. Springer, Berlin (1998)
4. Bart, K.: Neural Networks and Fuzzy Systems: A Dynamical Systems Approach to Machine Intelligence. Prentice-Hall, USA (1994)
5. Bede, B.: Mathematics of Fuzzy Sets and Fuzzy Logic. Springer, Berlin (2013)
6. Belohlávek, R.: Fuzzy Relational Systems: Foundations and Principles. Springer, Berlin (2002)
7. Benoet, R., Bruno, F., Philippe, D., Paul, H.J.: Vector Control of Induction Machines Desensitisation and Optimisation Through Fuzzy Logic (Power Systems) Vector Control of Induction Machines. Springer, Berlin (2012)
8. Benzaouia, A., Hajjaji, A.E.: Advanced Takagi-Sugeno Fuzzy Systems: Delay and Saturation. Springer, Berlin (2014)
9. Berenji, H.R., Khedkar, P.: Learning and tuning fuzzy logic controllers through reinforcements. IEEE Trans. Neural Netw. **3**, 724–740 (1992)
10. Bergmann, M.: An Introduction to Many-Valued and Fuzzy Logic: Semantics, Algebras, and Derivation Systems. Cambridge University Press, Cambridge (2008)
11. Bezdek, J.C., Keller, J., Krisnapuram, R., Pal, N.: Fuzzy Models and Algorithms for Pattern Recognition and Image Processing. Springer, Berlin (1999)
12. Bojadziev, G., Bojadziev, M.: Fuzzy Logic for Business, Finance, and Management. World Scientific Publishing Company, Singapore (2007)

13. Buckley, J.J.: Fuzzy Probability and Statistics. Springer, Berlin (2006)
14. Chadli, M., Borne, P.: Multiple Models Approach in Automation: Takagi-Sugeno Fuzzy Systems. Wiley, New Jersey (2012)
15. Chaira, T., Ray A.K.: Fuzzy Image Processing and Applications with MATLAB. CRC Press, USA (2009)
16. Chakraverty, S., Tapaswini, S., Behera, D.: Fuzzy Arbitrary Order System: Fuzzy Fractional Differential Equations and Applications. Wiley, New Jersey (2016)
17. Chen, G., Pham, T.T.: Introduction to Fuzzy Sets, Fuzzy Logic, and Fuzzy Control Systems. CRC Press, USA (2000)
18. Cirstea, M., Dinu, A., McCormick, M., Khor, J.G.: Neural and Fuzzy Logic Control of Drives and Power Systems. Newnes, Oxford (2002)
19. Cox, E.: Fuzzy Modeling and Genetic Algorithms for Data Mining and Exploration. Morgan Kaufmann, USA (2005)
20. Cox, E., O'Hagan, M.: The Fuzzy Systems Handbook, Second Edition: A Practitioner's Guide to Building, Using, and Maintaining Fuzzy Systems. Morgan Kaufmann, USA (1998)
21. Czogała, E., Łęski, J.: Fuzzy and Neuro-Fuzzy Intelligent Systems. Physica, The Netherlands (2012)
22. Deboeck, G.J.: Trading on the Edge: Neural, Genetic, and Fuzzy Systems for Chaotic Financial Markets. Wiley, New Jersey (1994)
23. De Silva, C.W.: Intelligent Control: Fuzzy Logic Applications. CRC Press, USA (1995)
24. Dompere, K.K.: Social Goal-Objective Formation, Democracy and National Interest: A Theory of Political Economy Under Fuzzy Rationality. Springer, Berlin (2014)
25. Downing, K.L.: Intelligence Emerging: Adaptivity and Search in Evolving Neural Systems. The MIT Press, Cambridge (2015)
26. Dubois, D., Prade, H.: Fuzzy Sets and Systems: Theory and Applications. Academic Press, USA (1980)
27. El-Hawary, M.E. (ed.): Electric Power Applications of Fuzzy Systems. Wiley-IEEE Press, New York (1998)
28. Farinwata, S.S., Filev, D.P., Langari, R. (eds.): Fuzzy Control: Synthesis and Analysis. Wiley, New Jersey (2000)
29. Fasel, D.: Fuzzy Data Warehousing for Performance Measurement: Concept and Implementation. Springer, Berlin (2014)
30. Feng, G.: Analysis and Synthesis of Fuzzy Control Systems: A Model-Based Approach. CRC Press, USA (2010)
31. Fodor, J.C., Roubens, M.R.: Fuzzy Preference Modelling and Multicriteria Decision Support. Springer, Berlin (2010)
32. Fuller, R.: Introduction to Neuro-Fuzzy Systems. Physica, The Netherlands (2000)
33. Genta, G., Genta, G.: Motor Vehicle Dynamics. World Scientific Publishing Company, Singapore (1997)
34. Gerla, G.: Fuzzy Logic: Mathematical Tools for Approximate Reasoning (Trends in Logic). Springer, Berlin (2001)
35. Gibilisco, M.B., Gowen, A.M., Albert, K.E., Mordeson, J.N., Wierman, M.J., Clark, T.D.: Fuzzy Social Choice Theory. Springer, Berlin (2014)
36. Glykas, M.: Fuzzy Cognitive Maps: Advances in Theory, Methodologies, Tools and Applications. Springer, Berlin (2010)
37. Gomes, L.T., de Barros, L.C., Bede, B.: Fuzzy Differential Equations in Various Approaches. Springer, Berlin (2015)
38. Gopal, M.: Digital Control and State Variable Methods: Conventional and Neuro-fuzzy Control Systems. McGraw Hill Higher Education, New York (2010)
39. Grint, K.: Fuzzy Management: Contemporary Ideas and Practices at Work. Oxford University Press, Oxford (1998)
40. Hopgood, A.A.: Intelligent Systems for Engineers and Scientists. CRC Press, USA (2000)
41. Höppner, F., Klawonn, F., Kruse, R., Runkler, T.: Fuzzy Cluster Analysis: Methods for Classification, Data Analysis and Image Recognition. Wiley, New Jersey (1999)

42. Ibrahim, A.: Fuzzy Logic for Embedded Systems Applications. Newnes, Oxford (2003)
43. Jager, T.: Black-Box Models of Computation in Cryptology. Vieweg+Teubner Verlag, Berlin (2012)
44. Jamshidi, M.: Large-Scale Systems: Modeling, Control and Fuzzy Logic. Prentice Hall, USA (1996)
45. Jang, J.S.R.: ANFIS: adaptive network based fuzzy inference system. IEEE Trans. Syst. Man Cybern. **123**, 665–685 (1993)
46. Jang, J.S.R., Sun, C.T., Mizutani, E.: Neuro-Fuzzy and Soft Computing: A Computational Approach to Learning and Machine Intelligence. Prentice Hall, USA (1997)
47. Jantzen, J.: Foundations of Fuzzy Control: A Practical Approach. Wiley, New Jersey (2013)
48. Jin, Y., Wang, L. (eds.): Fuzzy Systems in Bioinformatics and Computational Biology. Springer, Berlin (2010)
49. Kandel, A., Langholz, G.: Fuzzy Control Systems. CRC Press, USA (1993)
50. Kasabov, N.K.: Foundations of Neural Networks, Fuzzy Systems, and Knowledge Engineering. The MIT Press, Cambridge (1996)
51. Kasabov, N.K.: DENFIS: dynamic evolving neural-fuzzy inference system and its application for time series prediction. IEEE Trans. Fuzzy Syst. **10**, 144–154 (2002)
52. Kayacan, E., Khanesar, M.A.: Fuzzy Neural Networks for Real Time Control Applications: Concepts, Modeling and Algorithms for Fast Learning. Butterworth-Heinemann (2015)
53. Kecman, V.: Learning and Soft Computing: Support Vector Machines, Neural Networks, and Fuzzy Logic Models. A Bradford Book (2001)
54. Klir, G.J., Yuan, B.: Fuzzy Sets and Fuzzy Logic: Theory and Applications. Prentice Hall, USA (1995)
55. Klir, G.J., Clair, U.S., Yuan, B.: Fuzzy Set Theory: Foundations and Applications. Prentice Hall, USA (1997)
56. Kosko, B.: Neural Networks and Fuzzy Systems: A Dynamical Systems Approach to Machine Intelligence. Prentice Hall, USA (1991)
57. Kosko, B.: Fuzzy Thinking: The New Science of Fuzzy Logic. Hyperion, New York (1994)
58. Kruse, R., Gebhardt, J.E., Klawonn, F.: Foundations of Fuzzy Systems. Wiley, New Jersey (1994)
59. Kulkarni, A.D.: Computer Vision and Fuzzy-Neural Systems. Prentice Hall PTR, USA (2001)
60. Lewis, H.W.: The Foundations of Fuzzy Control. Springer, Berlin (1997)
61. Lilly, J.H.: Fuzzy Control and Identification. Wiley, New Jersey (2010)
62. Lin, C.T., Lee, C.S.G.: Neural network based fuzzy logic control and decision system. IEEE Trans. Comput. **40**, 1320–1336 (1991)
63. Lin, C.T., Lee, C.S.G.: Neural Fuzzy Systems: A Neuro-Fuzzy Synergism to Intelligent Systems. Prentice Hall, USA (1996)
64. Malik, D.S., Mordeson, J.N.: Fuzzy Discrete Structures. Physica, The Netherlands (2000)
65. Mamdani, E.H.: Application of fuzzy algorithms for control of simple dynamic plant. Proc. of IEEE **121**, 1585–1588 (1974)
66. Mamdani, E.H., Assilian, S.: An experiment in linguistic synthesis with a fuzzy logic controller. Int. J. Man-Mach. Stud. **7**, 1–13 (1975)
67. Massad, E., Ortega, N.R.S, de Barros, L.C., Struchiner, C.J.: Fuzzy Logic in Action: Applications in Epidemiology and Beyond (Studies in Fuzziness and Soft Computing). Springer, Berlin (2010)
68. McNeill, D., Freiberger, P.: Fuzzy Logic: The Revolutionary Computer Technology that is Changing our World. Simon & Schuste, New York (1994)
69. Melin, P., Castillo, O.: Design of Intelligent Systems Based on Fuzzy Logic, Neural Networks and Nature-Inspired Optimization. Springer, Berlin (2015)
70. Mendel, J.M.: Uncertain Rule-Based Fuzzy Logic Systems: Introduction and New Directions. Prentice Hall, USA (2001)
71. Mitra, S., Gupta, M.M., Kraske, W.F. (eds.): Neural and Fuzzy Systems: The Emerging Science of Intelligent Computing. Society of Photo Optical (1994)

72. Mordeson, J.N., Malik, D.S., Cheng, S.C.: Fuzzy Mathematics in Medicine. Physica, The Netherlands (2000)
73. Nanayakkara, T., Jamshidi, M., Sahin, F.: Intelligent Control Systems with an Introduction to System of Systems Engineering. CRC Press, USA (2009)
74. Nauck, D., Klawonn, F., Kruse, R.: Foundations of Neuro-Fuzzy Systems. Wiley, New Jersey (1997)
75. Nauck, D., Kruse, R.: Neuro-fuzzy systems for function approximation. Fuzzy Sets Syst. **101**, 261–271 (1999)
76. Nedjah, N., de Macedo Mourelle, L.: Fuzzy Systems Engineering: Theory and Practice. Springer, Berlin (2005)
77. Negoita, C.V.: Expert Systems and Fuzzy Systems. Benjamin-Cummings Pub. Co (1985)
78. Nelles, O.: Nonlinear System Identification: From Classical Approaches to Neural Networks and Fuzzy Models. Springer, Berlin (2000)
79. Nguyen, H.T., Sugeno, M. (eds.): Fuzzy Systems: Modeling and Control. Springer, Berlin (1998)
80. Nguyen, H.T., Walker, E.A.: A First Course in Fuzzy Logic. Chapman and Hall/CRC, UK (2005)
81. Nowicki, R.: On combining neuro-fuzzy architectures with the rough set theory to solve classification problems with incomplete data. IEEE Trans. Knowl. Data Eng. **39**, 1239–1253 (2008)
82. Nowicki, R.: Rough neuro-fuzzy structures for classification with missing data. IEEE Trans. Syst. Man Cybern. Part B **39**, 1334–1347 (2009)
83. Palm, R., Driankov, D., Hellendoorn, H.: Model Based Fuzzy Control: Fuzzy Gain Schedulers and Sliding Mode Fuzzy Controllers. Springer, Berlin (1997)
84. Passino, K.M., Yurkovich, S.: Fuzzy Control. Addison Wesley Publishing Company, USA (1997)
85. Pawar, P.M., Ganguli, R.: Structural Health Monitoring Using Genetic Fuzzy Systems. Springer, Berlin (2014)
86. Pedrycz, W., Gomide, F.: An Introduction to Fuzzy Sets: Analysis and Design. A Bradford Book (1998)
87. Pedrycz, W., Gomide, F.: Fuzzy Systems Engineering: Toward Human-Centric Computing. Wiley-IEEE Press, New York (2007)
88. Pedrycz, W., Ekel, P., Parreiras, R.: Fuzzy Multicriteria Decision-Making: Models, Methods and Applications. Wiley, New Jersey (2010)
89. Ponce-Cruz, P., Molina, A., MacCleery, B.: Fuzzy Logic Type 1 and Type 2 Based on LabVIEW(TM) FPGA. Springer, Berlin (2016)
90. Prade, H., Yager, R.R., Dubois, D. (eds.): Readings in Fuzzy Sets for Intelligent Systems. Morgan Kaufmann Publishers, USA (1993)
91. Principe, J.C., Euliano, N.R., Lefebvre, W.C.: Neural and Adaptive Systems: Fundamentals through Simulations. Wiley, New Jersey (1999)
92. Reznik, L.: Fuzzy Controllers Handbook: How to Design Them, How They Work. Newnes, Oxford (1997)
93. Reznik, L., Dimitrov, V. (eds.): Fuzzy Systems Design: Social and Engineering Applications. Physica, The Netherlands (2013)
94. Robyns, B., Francois, B., Degobert, P., Hautier, J.P.: Vector Control of Induction Machines: Desensitisation and Optimisation Through Fuzzy Logic. Springer, Berlin (2012)
95. Ruan, D. (ed.): Fuzzy Systems and Soft Computing in Nuclear Engineering. Physica, The Netherlands (2011)
96. Rutkowska, D.: Neuro-Fuzzy Architectures and Hybrid Learning. Springer, Berlin (2002)
97. Rutkowski, L.: Flexible Neuro-Fuzzy Systems: Structures, Learning and Performance Evaluation. Springer, Berlin (2004)
98. Rutkowski, L.: Computational Intelligence. Springer, Berlin (2010)
99. Sadeghian, A., Mendel, J., Tahayori, H. (eds.): Advances in Type-2 Fuzzy Sets and Systems: Theory and Applications. Springer, Berlin (2013)

100. Sanchez E.: Genetic Algorithms and Fuzzy Logic Systems Soft Computing Perspectives. Wspc, Singapore (1997)
101. Scherer, R.: Multiple Fuzzy Classification Systems. Springer, Berlin (2012)
102. Seising, R. (ed.): Views on Fuzzy Sets and Systems from Different Perspectives: Philosophy and Logic, Criticisms and Applications. Springer, Berlin (2010)
103. Shukla, S.K., Tiwari, A.K.: Efficient Algorithms for Discrete Wavelet Transform: With Applications to Denoising and Fuzzy Inference Systems. Springer, Berlin (2013)
104. Siddique, N., Adeli, H.: Computational Intelligence: Synergies of Fuzzy Logic, Neural Networks and Evolutionary Computing. Wiley, New Jersey (2013)
105. Siler, W., Buckley, J.J.: Fuzzy Expert Systems and Fuzzy Reasoning. Wiley, New Jersey (2004)
106. Slowinski, R. (ed.): Fuzzy Sets in Decision Analysis, Operations Research and Statistics. Springer, Berlin (1998)
107. Spooner, J.T., Maggiore, M., Ordóñez, R., Passino, K.M.: Stable Adaptive Control and Estimation for Nonlinear Systems: Neural and Fuzzy Approximator Techniques. Wiley, New Jersey (2001)
108. Starczewski, J.: Advanced Concepts in Fuzzy Logic and Systems with Membership Uncertainty. Springer, Berlin (2013)
109. Suykens, J.A.K., Vandewalle, J.P.L.: Nonlinear Modeling: Advanced Black-Box Techniques. Springer, Berlin (1998)
110. Syropoulos, A.: Theory of Fuzzy Computation. Springer, Berlin (2014)
111. Takagi, T., Sugeno, M.: Fuzzy identification of systems and its application to modeling and control. IEEE Trans. Syst. Man Cybern. **15**, 116–132 (1985)
112. Terano, T., Asa, K., Sugeno, M. (eds.): Fuzzy Systems Theory and its Applications. Academic Press, USA (1992)
113. Tirozzi, B., Puca, S., Pittalis S., Bruschi, A., Morucci, S., Ferraro, E., Corsini, S.: Neural Networks and Sea Time Series: Reconstruction and Extreme-Event Analysis. Birkhäuser, Switzerland (2005)
114. Tsoukalas, L.H., Uhrig, R.E., Zadeh, L.A.: Fuzzy and Neural Approaches in Engineering. Wiley, New Jersey (1997)
115. Tzeng, G.H., Huang, J.J.: Fuzzy Multiple Objective Decision Making. Chapman and Hall/CRC, UK (2013)
116. Von Altrock, C.: Fuzzy Logic and Neuro-Fuzzy Applications Explained. Prentice Hall, USA (1995)
117. Von Altrock, C.: Fuzzy Logic and Neuro-Fuzzy Applications in Business and Finance. Prentice Hall, USA (1996)
118. Vukadinovic, D. (ed.): Fuzzy Control Systems (Mathematics Research Developments). Nova Science Pub. Inc., New York (2011)
119. Wang, L.X.: A Course in Fuzzy Systems and Control. Prentice Hall, USA (1996)
120. Wang, Z.: Nonlinear Integrals and their Applications in Data Mining. World Scientific Publishing Company, Singapore (2010)
121. Wang, J.S., Lee, C.S.G.: Self adaptive neuro fuzzy inference systems for classification applications. IEEE Trans. Fuzzy Syst. **10**, 790–802 (2002)
122. Xu, J., Zen, Z.: Fuzzy-Like Multiple Objective Multistage Decision Making. Springer, Berlin (2014)
123. Yager, R.R., Zadeh, L.A.: An Introduction to Fuzzy Logic Applications in Intelligent Systems. Springer, Berlin (1992)
124. Yamakawa, T., Miki, T.: The current mode fuzzy logic integrated circuits fabricated by the standard CMOS process. IEEE Trans. Comput. **C-35**, 161–167 (1986)
125. Yen, J., Langari, R., Zadeh, L.A. (eds.): Industrial Applications of Fuzzy Logic and Intelligent Systems. Institute of Electrical (1995)
126. Ying-Ming, L.: Fuzzy Topology. World Scientific Publishing Company, Singapore (1998)
127. Zadeh, L.A.: Fuzzy sets. Inf. Control **8**, 338–353 (1965)

128. Zadeh, L., Kacprzyk, J.: Fuzzy Logic for the Management of Uncertainty. Wiley, New Jersey (1992)
129. Zavala, A.H., Nieto, O.C.: Fuzzy hardware: a retrospective and analysis. IEEE Trans. Fuzzy Syst. **20**, 623–635 (2012)
130. Zhang, H., Liu, D.: Fuzzy Modeling and Fuzzy Control. Birkhäuser, Switzerland (2006)
131. Zimmermann H.J.: Fuzzy Set Theory and its Applications. Springer, Berlin (2013)

Chapter 2
Selected Topics in Fuzzy Systems Designing

The research field of fuzzy systems is based on the theory of fuzzy sets [78]. However, it comprises a number of issues, which are related to, among others, triangular norms (see e.g. [2, 30, 31]), negations (see e.g. [59, 68]), inference operators (see e.g. [4, 5, 39, 44]), defuzzification methods (see e.g. [18, 34, 73]), a way of learning (see e.g. [54–56]), etc. For this reason, the scope of this chapter is limited and it describes the issues necessary for proper understanding of the contents presented in the subsequent chapters. The chapter contains a description of Mamdani-type and logical-type fuzzy systems (Sect. 2.1), an introduction to the learning methods of fuzzy systems (Sect. 2.2), and the evaluation methods of their performance when used in different application areas (Sect. 2.3).

2.1 Fuzzy System Description

This section provides a description of the fuzzy system which is used to present aspects of interpretable fuzzy systems designing. This system may have multiple inputs and multiple outputs (after the generalization of one-output system). Thus, it maps $\mathbf{X} \rightarrow \mathbf{Y}$, where input space data $\mathbf{X} = X_1, \ldots, X_n \subset \mathbf{R}^n$ and output space data $\mathbf{Y} = Y_1, \ldots, Y_m \subset \mathbf{R}^m$. The structure of the considered fuzzy system includes four main blocks: fuzzy rules base, inference block, fuzzification block and defuzzification block.

Fuzzification block performs a transformation (fuzzification operation) of not fuzzy (sharp) space $\mathbf{X} \subset \mathbf{R}^n$ to the space of fuzzy sets defined in $\mathbf{X}$. Due to fuzzification it is possible to give real number values (typical for most practical applications) at the system inputs. The most often realized fuzzification operation is so-called singleton fuzzification, which maps $\bar{\mathbf{x}} = [\bar{x}_1, \ldots, \bar{x}_n] \in \mathbf{X}$ to the fuzzy set $A' \subseteq \mathbf{X}$ which has the following membership function:

K. Cpałka, *Design of Interpretable Fuzzy Systems*, Studies in Computational Intelligence 684, DOI 10.1007/978-3-319-52881-6_2

$$\mu_{A'}(\mathbf{x}) = \begin{cases} 1 \text{ if } \mathbf{x} = \bar{\mathbf{x}} \\ 0 \text{ if } \mathbf{x} \neq \bar{\mathbf{x}} \end{cases}. \tag{2.1}$$

In addition to singleton fuzzification, there is also non-singleton fuzzification (see e.g. [27, 41, 54]). It can be used e.g. for filtering noise from input signals of a fuzzy system.

A fuzzy rules base contains the collection of N fuzzy rules. Each of them can be interpreted as a fuzzy relation on the set $\mathbf{X} \times \mathbf{Y}$. It has the following form:

$$R^k : \left[\begin{array}{c} \text{IF } \left(x_1 \text{ is } A_1^k\right) \text{AND} \ldots \text{AND } \left(x_n \text{ is } A_n^k\right) \\ \text{THEN } \left(y_1 \text{ is } B_1^k\right), \ldots, \left(y_m \text{ is } B_m^k\right) \end{array} \right], \tag{2.2}$$

where n is the number of input linguistic variables (system inputs), m is the number of output linguistic variables (system outputs), $\mathbf{x} = [x_1, \ldots, x_n] \in \mathbf{X}$ is the vector of input linguistic variables, $\mathbf{y} = [y_1, \ldots, y_m] \in \mathbf{Y}$ is the vector of output linguistic variables, $A_1^k, \ldots, A_n^k$ $(k = 1, \ldots, N)$ are input fuzzy sets representing values of input linguistic variables, $B_1^k, \ldots, B_m^k$ $(k = 1, \ldots, N)$ are output fuzzy sets representing values of output linguistic variables, $\mu_{A_i^k}(x_i)$ are membership functions of input fuzzy sets and $\mu_{B_j^k}(y_j)$ are membership functions of output fuzzy sets. Thus, linguistic variables do not have numeric values but the ones described by fuzzy sets. These sets represent descriptive notions like "low", "high", "close to the value of 5", etc. The notions cannot be directly processed e.g. in artificial neural-networks. Shape and distribution analysis of input and output fuzzy sets has an important impact on their interpretability.

The inference block processes input fuzzy values of the system and generates the output fuzzy value. First, fuzzy conclusions are determined from the system rules. They have a form of fuzzy sets $\bar{B}_j^k$, generated independently for each rule k and each output j of the system. When using the modus ponens generalized inference rule, which is the most commonly used [56], the set $\bar{B}_j^k$ has the following form:

$$\bar{B}_j^k = A' \circ \left(\mathbf{A}^k \rightarrow B_j^k\right), \tag{2.3}$$

where $\mathbf{A}^k \rightarrow B_j^k$ is the fuzzy relation represented by the fuzzy rule R^k and $\mathbf{A}^k = A_1^k \times \ldots \times A_n^k$ is a Cartesian product of fuzzy sets $A_1^k, \ldots, A_n^k$. Values of membership functions of fuzzy sets $\bar{B}_j^k$ of the form (2.3) are determined as follows:

$$\mu_{\bar{B}_j^k}(y_j) = \sup_{\mathbf{x} \in \mathbf{X}} \left\{ T \left\{ \mu_{A'}(\mathbf{x}), \mu_{\mathbf{A}^k \rightarrow B_j^k}(\mathbf{x}, y_j) \right\} \right\}, \tag{2.4}$$

where t-norm $T\{\cdot\}$ is a generalization of a conjunction operator known from a bivalent logic (see e.g. [2, 30, 31, 55, 56]). A set of sample t-norms is presented in Table 2.1. Using the singleton defuzzification (2.1) and a boundary condition for t-norm, the dependency (2.4) can be simplified to the following form:

Table 2.1 A list of exemplary triangular norms

Name	t-norm and t-conorm
Minimum/Maximum	$\begin{cases} T_m\{\mathbf{a}\} = \min\limits_{i=1,\dots,n}\{a_i\} \\ S_m\{\mathbf{a}\} = \max\limits_{i=1,\dots,n}\{a_i\} \end{cases}$
Łukasiewicz	$\begin{cases} T_L\{\mathbf{a}\} = \max\left\{\sum\limits_{i=1}^{n} a_i - (n-1), 0\right\} \\ S_L\{\mathbf{a}\} = \min\left\{\sum\limits_{i=1}^{n} a_i, 1\right\} \end{cases}$
Algebraic	$\begin{cases} T_a(\mathbf{a}) = \prod\limits_{i=1}^{n} a_i \\ S_a(\mathbf{a}) = 1 - \prod\limits_{i=1}^{n}(1-a_i) \end{cases}$
Drastic	$\begin{cases} T_d\{\mathbf{a}\} = \begin{cases} 0 & \text{for } S_m\{\mathbf{a}\} < 1 \\ T_m\{\mathbf{a}\} & \text{for } S_m\{\mathbf{a}\} = 1 \end{cases} \\ S_d\{\mathbf{a}\} = \begin{cases} 1 & \text{for } T_m\{\mathbf{a}\} > 0 \\ S_m\{\mathbf{a}\} & \text{for } T_m\{\mathbf{a}\} = 0 \end{cases} \end{cases}$

$$\mu_{\bar{B}_j^k}\left(y_j\right) = \mu_{\mathbf{A}^k \to B_j^k}\left(\bar{\mathbf{x}}, y_j\right) = I\left(\mu_{\mathbf{A}^k}\left(\bar{\mathbf{x}}\right), \mu_{B_j^k}\left(y_j\right)\right), \tag{2.5}$$

where $I(\cdot)$ is an inference operator depending on the inference type (see Sects. 2.1.1 and 2.1.2). Influencing the accuracy of inference operator used in Eq. (2.5) is an important aspect of influence on interpretability. Notation $\mathbf{A}^k$ in dependency (2.5) is interpreted as the activity level of the rule R^k and it is determined as follows:

$$\mu_{\mathbf{A}^k}\left(\bar{\mathbf{x}}\right) = T\left\{\mu_{A_1^k}\left(\bar{x}_1\right), \dots, \mu_{A_n^k}\left(\bar{x}_n\right)\right\} = \mathop{T}_{i=1}^{n}\left\{\mu_{A_i^k}\left(\bar{x}_i\right)\right\} = \tau_k\left(\bar{\mathbf{x}}\right). \tag{2.6}$$

Analysis of rules activity and influencing a precision of the aggregation operator (t-norm) used in Eq. (2.6) are an important aspect of influence on interpretability.

Next, conclusions from fuzzy rules $\bar{B}_j^k$ (their number is equal to $N \cdot m$) generated in the inference block are aggregated to the fuzzy conclusions from the whole rule base (their number is equal to m). They take the form of fuzzy sets B_j' $(j = 1, \dots, m)$. The sets are described by membership functions $\mu_{B_j'}\left(y_j\right)$. They are created as a result of aggregation of the sets $\bar{B}_j^k$ in a manner dependent on the assumed inference type (Sects. 2.1.1 and 2.1.2). In addition to the described implementation of the inference block, there are also other ways based e.g. on another inference rule (e.g. the modus tollens generalized inference rule), another way of generating conclusions from the fuzzy rules base, etc. (see e.g. [56]).

The defuzzification block performs the inverse operation of the one implemented in the fuzzification block. It provides a transition from the fuzzy sets B_j'

$(j = 1, \ldots, m)$ obtained on the output of the inference block to the real (sharp) output signals $\bar{y}_j$ from the fuzzy system in the space $\mathbf{Y} \subset \mathbf{R}$. In practice, there are many different defuzzification operators (see e.g. [18, 34, 73]) and new ones are still being developed. For the purpose of further consideration we have assumed that the inference will be implemented by the commonly used center-of-area method. It has the following form:

$$\bar{y}_j = \frac{\int\limits_{\mathbf{Y}} y_j \cdot \mu_{B'_j}\left(y_j\right) dy_j}{\int\limits_{\mathbf{Y}} \mu_{B'_j}\left(y_j\right) dy_j}. \tag{2.7}$$

The discrete form of dependency (2.7) is usually written as follows:

$$\bar{y}_j = \frac{\sum\limits_{r=1}^{N} \bar{y}_{j,r}^B \cdot \mu_{B'_j}\left(\bar{y}_{j,r}^B\right)}{\sum\limits_{r=1}^{N} \mu_{B'_j}\left(\bar{y}_{j,r}^B\right)}, \tag{2.8}$$

where $\bar{y}_{j,r}^B$ $(r = 1, \ldots, N,\ j = 1, \ldots, m)$ are discretization points of fuzzy set B'_j. Most often they are the points at which input fuzzy sets B_j^k reach their maximum values. If fuzzy sets B_j^k reach their maximum in a range value, then, another defuzzification method can be used. However, it should be noted that most of the defuzzification methods is dependent on the number N of the system rules. It means a negative dependency between precision of defuzzification operator and the number of fuzzy rules. An analysis of this issue is an important aspect of affecting interpretability.

2.1.1 *Fuzzy System Implementing Mamdani-Type Inference*

Specifics of the Mamdani-type fuzzy system result from an inference block implementation method. In this system a t-norm is most often the inference operator. However, attempts are also being made to use other operators which meet some or all assumptions arising from the t-norm definition (see e.g. [31, 38, 44]). In the Mamdani-type system the dependency (2.5) can be written as follows:

$$I\left(\mu_{\mathbf{A}^k}\left(\bar{\mathbf{x}}\right), \mu_{B_j^k}\left(y_j\right)\right) = T\left\{\mu_{\mathbf{A}^k}\left(\bar{\mathbf{x}}\right), \mu_{B_j^k}\left(y_j\right)\right\}. \tag{2.9}$$

In turn, aggregation of fuzzy conclusions $\bar{B}_j^k$ from fuzzy rules R^k to final conclusions B'_j $(j = 1, \ldots, m)$ is performed as follows:

$$B'_j = \bigcup_{k=1}^{N} \bar{B}_j^k. \tag{2.10}$$

Membership functions of fuzzy sets B'_j are determined using the following formula:

$$\mu_{B'_j}\left(y_j\right) = S\left\{\mu_{\bar{B}^1_j}\left(y_j\right),\ldots,\mu_{\bar{B}^N_j}\left(y_j\right)\right\} = \underset{k=1}{\overset{N}{S}}\left\{\mu_{\bar{B}^k_j}\left(y_j\right)\right\}, \tag{2.11}$$

where t-conorm $S\{\cdot\}$ is a generalization of an alternative operator known from a bivalent logic (see e.g. [2, 30, 31, 55, 56]). It means that the rules of the form (2.2) are associated by the operator which is an extension of the OR operator. A set of sample t-conorms is presented in Table 2.1.

Taking into account formulas (2.6), (2.9) and (2.11) in the assumed defuzzification formula (2.8), a dependency describing the value of the output signal $\bar{y}_j$ for the Mamdani-type system is obtained:

$$\bar{y}_j = \frac{\sum\limits_{r=1}^{N} \bar{y}^B_{j,r} \cdot \underset{k=1}{\overset{N}{S}}\left\{T\left\{\underset{i=1}{\overset{n}{T}}\left\{\mu_{A^k_i}\left(\bar{x}_i\right)\right\}, \mu_{B^k_j}\left(\bar{y}^B_{j,r}\right)\right\}\right\}}{\sum\limits_{r=1}^{N} \underset{k=1}{\overset{N}{S}}\left\{T\left\{\underset{i=1}{\overset{n}{T}}\left\{\mu_{A^k_i}\left(\bar{x}_i\right)\right\}, \mu_{B^k_j}\left(\bar{y}^B_{j,r}\right)\right\}\right\}}. \tag{2.12}$$

A system implementing an alternative inference type is a logical-type system. It is described in the next section.

2.1.2 Fuzzy System Implementing Logical-Type Inference

Specifics of the logical-type fuzzy system also (like in the Mamdani-type system) result from an inference block implementation method. In this system the inference operator may be e.g.: s-implication, r-implication, q-implication, or other operators which are usually an extension of logical implication (see e.g. [4, 5, 31, 38, 44]). In this book systems using s-implication inference operator are considered. Then, the dependency (2.5) can be presented as follows:

$$I\left(\mu_{\mathbf{A}^k}\left(\bar{\mathbf{x}}\right), \mu_{B^k_j}\left(y_j\right)\right) = S\left\{\operatorname{neg}\left(\mu_{\mathbf{A}^k}\left(\bar{\mathbf{x}}\right)\right), \mu_{B^k_j}\left(y_j\right)\right\}, \tag{2.13}$$

where neg $(\cdot)$ is a negation operator (see e.g. [31, 44, 59, 68]). There are many different operators consistent with the definition of the negation (e.g. Zadeh, Hamacher, Sugeno, Dombi, Yager, etc.), but in practice Zadeh negation is actually the one which is most often used:

$$\operatorname{neg}\left(\mu_{\mathbf{A}^k}\left(\bar{\mathbf{x}}\right)\right) = 1 - \mu_{\mathbf{A}^k}\left(\bar{\mathbf{x}}\right). \tag{2.14}$$

In the considered system of the logical-type an aggregation of fuzzy conclusions $\bar{B}^k_j$ from fuzzy rules R^k to final conclusions B'_j $(j = 1,\ldots,m)$ is performed as follows:

$$B'_j = \bigcap_{k=1}^{N} \bar{B}^k_j. \tag{2.15}$$

Membership functions of fuzzy sets B'_j are determined using the following formula:

$$\mu_{B'_j}(y_j) = T\left\{\mu_{\bar{B}^1_j}(y_j), \ldots, \mu_{\bar{B}^N_j}(y_j)\right\} = \overset{N}{\underset{k=1}{T}}\left\{\mu_{\bar{B}^k_j}(y_j)\right\}. \tag{2.16}$$

It means that the rules of the form (2.2) are associated by the operator which is an extension of the AND operator.

Taking into account formulas (2.6), (2.13), (2.14) and (2.16) in the assumed defuzzification formula (2.8), a dependency describing the value of the output signal $\bar{y}_j$ for the logical-type system is obtained:

$$\bar{y}_j = \frac{\sum_{r=1}^{N} \bar{y}^B_{j,r} \cdot \overset{N}{\underset{k=1}{T}}\left\{S\left\{1 - \overset{n}{\underset{i=1}{T}}\left\{\mu_{A^k_i}(\bar{x}_i)\right\}, \mu_{B^k_j}\left(\bar{y}^B_{j,r}\right)\right\}\right\}}{\sum_{r=1}^{N} \overset{N}{\underset{k=1}{T}}\left\{S\left\{1 - \overset{n}{\underset{i=1}{T}}\left\{\mu_{A^k_i}(\bar{x}_i)\right\}, \mu_{B^k_j}\left(\bar{y}^B_{j,r}\right)\right\}\right\}}. \tag{2.17}$$

Systems of the forms (2.12) and (2.17) are the base systems for the considerations presented in Chap. 4.

2.2 Fuzzy System Learning Methods

In practice, there are several common approaches to determining the structure and parameters of fuzzy systems:

- The first one assumes that the fuzzy rules and their fuzzy sets are formulated by an expert. It determines the structure and parameters of a system. If indications of the expert are accurate, then the learning of such fuzzy system is not needed.
- The second approach assumes that we have the expert knowledge (as it is in the first approach) and a learning sequence containing learning sets. Each set contains input signals of the system and corresponding output reference signals. Then, the system initialized by the expert can be also learned (tuned), thus increasing the precision of its operation.
- The third approach (the most common) assumes that we have only learning data at our disposal. They are the basis for learning of the fuzzy system. The learning may be unsupervised or/and supervised (see e.g. [47, 48]). In the most basic applications the purpose of learning is selection of parameters of the fuzzy system whose structure has been specified by the designer (e.g. the number of rules determined by trial and error). The purpose of learning is also increasingly more often an automatic selection of the fuzzy system structure taking into account, among

others, additional requirements. They may include e.g. the degree of complexity, aspects of readability, etc. This approach is considered in Chaps. 4–7.
- The fourth approach also assumes that we only have learning data at our disposal. However, they are not used for iterative learning of the system. They are used in order to generate appropriate descriptors, which have a precise interpretation (purpose) in the fuzzy system. This approach is considered in Chap. 8.

As already mentioned, the issue of fuzzy systems learning is based on capabilities of unsupervised and supervised learning. Unsupervised learning can use data clustering techniques (see e.g. [1, 20, 72]). They are one of the oldest and most popular data mining methods. Their aim is to extract classes in a data set and associate them with objects on the basis of an assumed similarity function. Use of data clustering in fuzzy systems learning allows us, among others, to properly initiate parameters of the system and outline its structure (e.g. the number of fuzzy rules). Here the use of a cluster validity index is often found helpful. An example of using unsupervised learning for selection (initialization) of the fuzzy system structure is shown in Chap. 5.

However, the learning of fuzzy systems is mainly performed using supervised learning, which is based on two types of algorithms:

- Gradient algorithms. The purpose of the gradient algorithm is iterative selection of the direction of optimized objective function changes (in particular, the evaluation function of the fuzzy system) in order to find its extreme (usually the minimum). Determination of the direction of change for the next step is based on the knowledge of the optimized objective function gradient in the current step. The direction of change being sought is the one in which the optimized objective function changes with the greatest intensity (decreases or increases). The most common gradient methods used in practice include the error backpropagation method [11, 45, 69, 70], its variant with the so-called momentum term [52, 76, 77], the Levenberg–Marquardt method [33, 37, 75], least squares method [67], the Fletcher-Reeves conjugate gradient method [22], sub-gradient methods [6], etc.
- Population algorithms. These algorithms are heuristic methods and they process a population of solutions in order to effectively search a given space of considerations (which takes into account the range of searched parameters). A single solution of a population is called an individual or a chromosome (in the case of genetic algorithms). Such individual may be e.g. a suitably encoded system described by formula (2.8). It is evaluated by the evaluation function (fitness function). Its form is dependent on a given problem. The value of the evaluation function of an individual determines its chances of survival in a given population. In the case of fuzzy systems, not only can the accuracy of the system be evaluated, but also e.g. the complexity of its structure and readability of its rules.

At present, the topics related to population algorithms are developing intensively. The reasons for that include high flexibility of this class of methods in configuration of a population and also ways of its processing, which can result in various approaches to their division. Therefore, taking into account the issue of configuration of population, these algorithms can be divided into:

- Single-population algorithms. They use a single population of individuals. A generalized scheme of operation of such algorithms includes the following steps: (a) Initiation of individuals in a population. (b) Evaluation of individuals in the given population using a defined fitness function. (c) Modification of individuals according to the specifics of the algorithm. (d) Evaluation of new individuals in the population. (e) Replacing the old population with the new population of individuals according to the specifics of the algorithm (this step can be omitted for those algorithms which do not create new populations, but only modify the parameters of an existing population). (f) Checking the stopping criterion, which may take into account e.g. the execution of a certain number of steps or getting a satisfactory solution in the form of an adopted evaluation function. In the case where this condition is not met, the algorithm returns to step c. However, if the stop condition is satisfied, then the presentation of the best individual is performed and the algorithm terminates. A single sequence of steps c–f is called a single iteration of the algorithm or a single evolution.
- Multi-population algorithms. Operation scheme of multi-population algorithms is analogous to the single-population algorithms. The difference results from the need to manage multiple populations. Each component population can be learned individually. Later those populations can either compete or cooperate. In Table 2.2 typical interactions between populations determining specificity of the algorithm are shown. In multi-population algorithms there is also the phenomenon of migration, which is the transfer of individuals between component populations. The way of migration can be determined by the result of competition. Common migration topologies are shown in Fig. 2.1. The use of many populations is implemented so

Table 2.2 Summary of possible interactions that can occur between populations

Type of interaction	Impact on		Description of relation
	Population 1	Population 2	
Neutralism	None	None	Populations are independent
Mutualism	Positive	Positive	Mutual benefit
Commensalism	Positive	None	One-sided benefit
Competition	Negative	Negative	Both sides are negatively affected
Predation	Positive	Negative	One-sided benefits, while the other is negatively affected
Parasitism	Positive	Negative	One-sided benefits, while the other is negatively affected

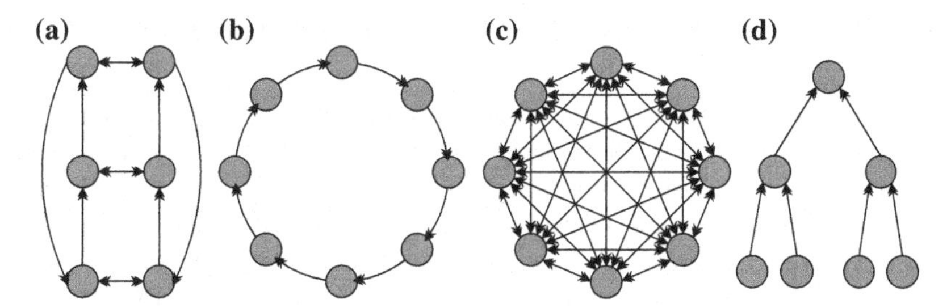

Fig. 2.1 Typical migration topologies between populations (the *arrows* indicate direction of migration): **a** ladder-type topology [9], **b** one-direction circle topology [8], **c** peer-to-peer topology [8], **d** hierarchical topology [49]

as to increase diversity of individuals in component populations and to improve the resistance of the algorithm to (negative) search for optimal solutions only locally.

Another criterion for division of population algorithms can be the way how a population is processed. In this context, population algorithms may be categorized, among others, as follows:

- Evolutionary algorithms. This group of algorithms can include all population-based algorithms which use the processing of population individuals based on biological evolution. The most commonly used evolutionary algorithms include genetic algorithms [24], evolution strategies [7], genetic programming [14, 51], evolutionary programming [23], differential evolution algorithms [21, 62–64], multi-population genetic algorithms [42], etc.
- Swarm algorithms. This group of algorithms may include all population-based algorithms which use processing of population individuals based on swarm behavior. The most commonly used swarm algorithms comprise the particle swarm optimization algorithm [12, 50], the artificial bee colony algorithm [29, 74], the intelligent water drops algorithm [60], the artificial immune algorithm [17, 65], the multi-swarm cooperative particle swarm optimization algorithm [46], etc.
- Other population-based algorithms. This group of algorithms can include all population-based algorithms which are difficult to define as evolutionary or swarm ones. They are algorithms based on league functioning (e.g. the golden ball algorithm [49]), biogeography algorithms (e.g. biogeography-based optimization algorithm [61]), algorithms based on colonization (e.g. invasive weed optimization [35]), algorithms referring to ways of spreading sparks (e.g. fireworks optimization algorithm [66]), algorithms modelling social revolution (e.g. multi-population imperialist competitive algorithm [3]), algorithms based on symbiosis (e.g. Particle Swarm^2 Optimization, PS^{2}0 [32]), etc.

From a practical point of view population algorithms can be grouped in a number of ways, but this, however, is not essential from the point of view of interpretable fuzzy systems design issues.

2.3 Methods for Evaluation of Fuzzy System Accuracy

Evaluation of fuzzy system accuracy depends on the field of application and formulated requirements (problem domain). It is implemented differently in classification issues and in regression (modelling) issues. Later in this chapter we describe basic ways used to evaluate performance of fuzzy systems used for classification and modelling. Moreover, it is pointed out that assessment of fuzzy systems accuracy in the issue of designing interpretable fuzzy systems is only one of the required/necessary/most often applied criteria for their overall evaluation.

2.3.1 *Accuracy Evaluation of the Systems Used for Modelling*

Accuracy evaluation of the systems used for modelling is most often performed using the root mean squared error (RMSE), which is defined as follows:

$$\mathrm{RMSE}\,(\mathbf{fs}) = \frac{1}{m} \cdot \sum_{j=1}^{m} \sqrt{\frac{1}{Z} \cdot \sum_{z=1}^{Z} \left(d_{z,j}^{L} - \bar{y}_{z,j} \right)^{2}}, \tag{2.18}$$

where **fs** represents any fuzzy system being evaluated (especially a system described by Eq. (2.8)), m is the number of the system outputs, Z is the number of samples from a learning or testing sequence, $d_{z,j}^{L}$ is an expected output value for output j $(j = 1, \ldots, m)$ for input vector z $(z = 1, \ldots, Z)$, $\bar{y}_{z,j}$ is a real output value j $(j = 1, \ldots, m)$ computed by the system for input vector z $(z = 1, \ldots, Z)$.

Since formula (2.18) does not take into account the disparity between the values of different outputs of a multiple input multiple output (MIMO) system, in this book a normalized error measure is also used and it has the following form:

$$\mathrm{acc}\,(\mathbf{fs}) = \frac{1}{m} \cdot \sum_{j=1}^{m} \frac{\frac{1}{Z} \cdot \sum_{z=1}^{Z} \left| d_{z,j}^{L} - \bar{y}_{z,j} \right|}{\max\limits_{z=1,\ldots,Z} \left\{ d_{z,j}^{L} \right\} - \min\limits_{z=1,\ldots,Z} \left\{ d_{z,j}^{L} \right\}}. \tag{2.19}$$

2.3.2 *Accuracy Evaluation of the Systems Used for Classification*

Fuzzy systems used for classification are characterized by various factors including the number of outputs, which is equal to the number of considered classes. In this case the input set belongs to a class associated with the system output whose signal value is the highest. Therefore, accuracy evaluation of the systems used for classification is performed using a standard classification error, which is determined as follows:

$$\text{acc}\,(\mathbf{fs}) = 100\% \cdot \frac{1}{Z}\sum_{z=1}^{Z}\begin{cases}1 \text{ for } yi_z \neq di_z\\ 0 \text{ for } yi_z = di_z\end{cases}, \tag{2.20}$$

where di_z is a reference index of the class for sample z ($z = 1, \ldots, Z$) and yi_z is an index of the class selected by the system for sample z ($z = 1, \ldots, Z$).

It should be noted that the 10-fold cross validation [71] is used most often for evaluation of the systems used for classification. In the 10-fold cross validation the learning sequence is divided into ten parts of equal size. Then, in the first step of this procedure the first part is treated as a testing sequence and the other 9 parts are treated as a learning sequence. In the next step of the procedure the second part becomes the testing sequence and the other ones act as the learning sequence, etc. Thus, the process is repeated ten times and the final result is determined by averaging the obtained partial results. In cross-validation procedure we often seek to preserve proportion between the number of class representatives which is similar to the one in the whole learning sequence. This is called stratified cross-validation.

2.3.3 Accuracy Evaluation in the Context of Interpretable Fuzzy Systems Designing

In interpretable fuzzy systems design issues system accuracy is only one of the evaluation criteria applied. The other criteria comprise complexity and interpretability. For this reason, solutions derived from multi-objective optimization [13, 15, 19, 25, 26, 36, 53] play an important role in the evaluation procedure. Multi-objective optimization allows us to take into account many components in the evaluation function. As a result, it is well suited for the evaluation of fuzzy systems in the context of, among others, accuracy, complexity and readability of rules. When considering the criteria for fuzzy system evaluation we may also affect their proper hierarchy of importance. Multi-criteria optimization methods also include:

- Scalar methods. They rely on aggregation of multiple criteria and require a priori knowledge. They include aggregation methods (e.g. random distribution of weights within a population [28]), methods based on LP-metrics [57], methods of ε-constraint type (i.e. a set of methods imposing restrictions on each evaluation function component [40]) as well as purpose methods (goal programming), in which the values of evaluation function components are optimized to certain threshold values [10].
- Methods based on the Pareto front. The front is a set of non-dominated solutions. They are the solutions for which there is no other better solution in terms of any criterion. These methods include ranking methods (based on a ranking of solutions,

e.g. NSGA-II [16]), elite methods (based only on non-dominated solutions, e.g. SEEA [80]) and methods managing the division (diversity maintaining [79]).
- Other methods. They include, among others, multi-population methods, methods based on interactions between populations (e.g. ERMOCS [43]) and methods with so-called parallel selection, which create vectors ordering sets of solutions in respect to each criterion independently (e.g. VEGA [58]).

Multi-criteria optimization techniques are perfectly suited to be used with population-based algorithms in the context of interpretable fuzzy system designing. These solutions are considered in Chap. 6.

2.4 Summary

In this chapter some selected issues in the field of fuzzy systems are outlined. Out of necessity the chapter focuses on the issues which are crucial for interpretable fuzzy systems designing. In particular, two types of fuzzy systems are described, i.e. the system implementing Mamdani-type inference and the system implementing logical-type inference. It is also emphasized that the considerations presented in this book have a general nature and they are not only related to the described class of systems. The presented description indicates the issues which have particular relevance to the subject matter of this book. Further on the chapter contains a review of basic techniques of fuzzy systems learning and methods for their evaluation. It discusses the aspects of supervised and unsupervised learning in the designing of fuzzy systems. Moreover, it emphasizes the importance of association of multi-objective optimization and evolutionary algorithms. The final part of the chapter indicates that accuracy of fuzzy systems is only one of the possible criteria for their evaluation.

References

1. Aggarwal, C.C., Reddy, C.K.: Data Clustering: Algorithms and Applications. Chapman and Hall/CRC, New York (2013)
2. Alsina, C., Maurice, F., Schweizer, B.: Associative Functions: Triangular Norms and Copulas. WSPC, New Jersey (2006)
3. Atashpaz–Gargari, E., Lucas, C.: Imperialist competitive algorithm: an algorithm for optimization inspired by imperialistic competition. In: Proceedings of the IEEE Congress on Evolutionary Computation, vol. 7, pp. 4661–4666 (2007)
4. Baczyński, M.: Fuzzy Implications. Springer, Heidelberg (2008)
5. Baczyński, M., Beliakov, G., Sola, H.B., Pradera, A. (eds.): Advances in Fuzzy Implication Functions. Springer, New York (2015)
6. Bertsekas, D.P.: Incremental proximal methods for large scale convex optimization. Math. Progr. **129**, 163–195 (2011)
7. Beyer, H.G., Schwefel, H.P.: Evolution strategies: a comprehensive introduction. Nat. Comput. **1**, 3–52 (2002)

8. Borovska, P., Lazarova, M.: Migration policies for Island genetic models on multicomputer platform. In: Intelligent Data Acquisition and Advanced Computing Systems: Technology and Applications (IDAACS 2007), pp. 143–148 (2007)
9. Cantu-Paz, E.: Topologies, migration rates, and multi-population parallel genetic algorithms. In: Proceedings of the Genetic and Evolutionary Computation Conference (GECCO 1999), pp. 91–98 (1999)
10. Charnes, A., Cooper, W.W., Ferguson, R.: Optimal estimation of executive compensation by linear programming. Manag. Sci. **1**, 138–151 (1955)
11. Chauvin, Y., Rumelhart, D.E. (eds.): Backpropagation: Theory, Architectures, and Applications. Psychology Press (2013)
12. Clerc, M.: Particle Swarm Optimization. Wiley-ISTE, London (2006)
13. Collette, Y., Siarry, P.: Multiobjective Optimization: Principles and Case Studies. Springer, New York (2011)
14. Cramer, N.L.: A representation for the adaptive generation of simple sequential programs. In: Proceedings of the International Conference on Genetic Algorithms and their Applications, pp. 183–187 (1985)
15. Deb, K.: Multi-Objective Optimization Using Evolutionary Algorithms. Wiley, New York (2009)
16. Deb, K., Pratap, A., Agarwal, S., Meyarivan, T.: A fast and elitist multiobjective genetic algorithm: NSGA-II. IEEE Trans. Evol. Comput. **6**, 182–197 (2002)
17. De Castro, L.N.: Artificial Immune Systems: A New Computational Intelligence Approach. Springer, New York (2002)
18. De Oliveira, J.V.: A set-theoretical defuzzification method. Fuzzy Sets Syst. **76**, 63–71 (1995)
19. Donoso, Y., Fabregat, R.: Multi-Objective Optimization in Computer Networks Using Metaheuristics. Auerbach Publications, Boca Raton (2007)
20. Everitt, B.S., Landau, S., Leese, M., Stahl, D.: Cluster Analysis. Wiley, New York (2011)
21. Feoktistov, V.: Differential Evolution: In Search of Solutions. Springer, New York (2006)
22. Fletcher, R., Reeves, C.: Function minimization by conjugate gradients. Comput. J. **7**, 149–154 (1964)
23. Fogel, L.J., Owens, A.J., Walsh, M.J.: Artificial Intelligence through Simulated Evolution. Wiley, New York (1966)
24. Fraser, A., Burnell, D.: Computer Models in Genetics. McGraw-Hill, New York (1970)
25. Gorzałczany, M., Rudziński, F.: A multi-objective genetic optimization for fast, fuzzy rule-based credit classification with balanced accuracy and interpretability. Appl. Soft Comput. **40**, 206–220 (2016)
26. Gorzałczany, M.B., Rudziński, F.: Accuracy versus interpretability of fuzzy rule-based classifiers: an evolutionary approach. In: Proceedings of the International Conference on Swarm and Evolutionary Computation (SIDE'12), pp. 222–230 (2012)
27. Heng, S., Chen, W., Yin, H., Shiping, M., Jizhang, Z.: Decision feedback equalizer based on non-singleton fuzzy regular neural networks. J. Syst. Eng. Electron. **17**, 896–900 (2006)
28. Jin, Y., Okabe, T., Sendhoff, B.: Adapting weighted aggregation for multiobjective evolutionary strategies. LNCS 1993, vol. 1993, pp. 96–110. Springer, Heidelberg (2006)
29. Karaboga, D.: An idea based on honey bee swarm for numerical optimization. Technical report-tr06. Erciyes University (2005)
30. Klement, E.P., Mesiar, R. (eds.): Logical, Algebraic, Analytic and Probabilistic Aspects of Triangular Norms. Elsevier, Amsterdam (2005)
31. Klement, E.P., Mesiar, R., Pap, E.: Triangular Norms. Springer, New York (2000)
32. Lenin, M.K., Reddy, B.R.: PS2O, hybrid evolutionary-conventional algorithm, genetical swarm optimization for solving reactive power optimization problem. Int. J. Recent Res. Electr. Electron. Eng. (IJRREEE) **1**, 10–20 (2014)
33. Levenberg, K.: A method for the solution of certain non-linear problems in least squares. Quart. Appl. Math. **2**, 164–168 (1944)
34. Mahdiani, H.R., Banaiyan, A., Javadi, M.H.S., Fakhraie, S.M., Lucas, C.: Defuzzification block: new algorithms, and efficient hardware and software implementation issues. Eng. Appl. Artif. Intell. **26**, 162–172 (2013)

35. Mallahzadeh, A.R.R., Oraizi, H., Davoodi–Rad, Z.: Application of the invasive weed optimization technique for antenna configurations. Prog. Electromagn. Res. **79**, 137–150 (2008)
36. Marler, T.: Multi-Objective Optimization: Concepts and Methods for Engineering. VDM Verlag, Germany (2009)
37. Marquardt, D.W.: An algorithm for least-squares estimation of nonlinear parameters. J. Soc. Ind. Appl. Math. **11**, 431–441 (1963)
38. Mas, M., Monserrat, M., Ruiz-Aguilera, D., Torrens, J.: RU and (U, N)-implications satisfying Modus Ponens. Int. J. Approx. Reason. (2016). doi:10.1016/j.ijar.2016.01.003
39. Mas, M., Monserrat, M., Torrens, J., Trillas, E.: A survey on fuzzy implication functions. IEEE Trans. Fuzzy Syst. **15**, 1107–1121 (2007)
40. Mavrotas, G.: Effective implementation of the e-constraint method in multi-objective mathematical programming problems. Appl. Math. Comput. **213**, 455–465 (2009)
41. Mouzouris, G.C., Mendel, J.M.: Dynamic non-singleton fuzzy logic systems for nonlinear modeling. IEEE Trans. Fuzzy Syst. **5**, 199–208 (1997)
42. Murata, T., Ishibuchi, H.: MOGA: multi-objective genetic algorithms. In: proceedings of the IEEE International Conference on Evolutionary Computation, vol. 102, pp. 289–294 (1995)
43. Neef, M., Thierens, D., Arciszewski, H.: A case study of a multiobjective elitist recombinative genetic algorithm with coevolutionary sharing. In: Proceedings of the 1999 Congress on Evolutionary Computation, pp. 796–803 (1999)
44. Nguyen, H.T., Walker, E.A.: A First Course in Fuzzy Logic. Chapman and Hall/CRC, Boca Raton (2005)
45. Nikolaev, N., Iba, H.: Adaptive Learning of Polynomial Networks: Genetic Programming, Backpropagation and Bayesian Methods. Springer, Boston (2006)
46. Niu, B., Zhu, Y., He, X., Wu, H.: MCPSO: a multi-swarm cooperative particle swarm optimizer. Appl. Math. Comput. **185**, 1050–1062 (2007)
47. Okun, O.: Supervised and Unsupervised Ensemble Methods and their Applications. Springer, Heidelberg (2008)
48. Okun, O.: Applications of Supervised and Unsupervised Ensemble Methods. Springer, Heidelberg (2010)
49. Osaba, E.: Golden ball: a novel meta-heuristic to solve combinatorial optimization problems based on soccer concepts. Appl. Intell. **12**, 145–166 (2013)
50. Parsopoulos, K.E., Vrahatis, M.N.: Particle Swarm Optimization and Intelligence: Advances and Applications. IGI Global, Pennsylvania (2010)
51. Poli, R., Langdon, W.B., McPhee, N.F.: A Field Guide to Genetic Programming. Lulu Enterprises (2008)
52. Qiu, G., Varley, M.R., Terrell, T.J.: Accelerated training of backpropagation networks by using adaptive momentum step. Electron. Lett. **28**, 377–379 (1992)
53. Rudziński, F.: A multi-objective genetic optimization of interpretability-oriented fuzzy rule-based classifiers. Appl. Soft Comput. **38**, 118–133 (2016)
54. Rutkowska, D.: Neuro-Fuzzy Architectures and Hybrid Learning. Springer, Heidelberg (2002)
55. Rutkowski, L.: Flexible Neuro-Fuzzy Systems: Structures, Learning and Performance Evaluation. Springer, Heidelberg (2004)
56. Rutkowski, L.: Computational Intelligence. Springer, Heidelberg (2010)
57. Sadjadi, S.J., Habibian, M., Khaledi, V.: A multi-objective decision making approach for solving quadratic multiple response surface problems. Int. J. Contemp. Math. Sci. **3**, 1595–1606 (2008)
58. Schaffer, J.: Multiple objective optimization with vector evaluated genetic algorithms. In: Proceedings of the First International Conference on Genetic Algorithms, pp. 93–100 (1985)
59. Schneider, M., Shnaider, E., Kandel, A.: Applications of the negation operator in fuzzy production rules. Fuzzy Sets Syst. **34**, 293–299 (1990)
60. Shah-Hosseini, H.: The intelligent water drops algorithm: a nature-inspired swarm-based optimization algorithm. Int. J. Bio-Inspir. Comput. **1**, 71–79 (2009)
61. Simon, D.: Biogeography-based optimization. IEEE Trans. Evol. Comput. **12**, 702–713 (2008)
62. Simon, D.: Evolutionary Optimization Algorithms. Wiley, New Jersey (2013)

63. Słowik, A.: Application of geometric differential evolution algorithm to design minimal phase digital filters with a typical characteristics for their hardware or software implementation. In: Proceedings of the 12th International Conference on Artificial Intelligence and Soft Computing, vol. 7895, pp. 67–78 (2013)
64. Storn, R.: On the usage of differential evolution for function optimization. In: Proceedings of the Biennial Conference of the North American Fuzzy Information Processing Society (NAFIPS), pp. 519–523 (1996)
65. Tan, Y.: Artificial Immune System: Applications in Computer Security. IEEE Computer Society Press Systems Series, Wiley (2016)
66. Tan, Y., Zhu, Y.: Fireworks algorithm for optimization. In: Proceedings of the First International Advances in Swarm Intelligence (ICSI2010). LNCS, vol. 6145, pp. 355–364. Springer, Heidelberg (2010)
67. Teunissen, P.J.G.: Dynamic Data Processing: Recursive Least-squares. VSSD, Netherlands (2009)
68. Weber, S.: A general concept of fuzzy connectives, negations and implications based on t-norms and t-conorms. Fuzzy Sets Syst. **11**, 115–134 (1983)
69. Werbos, P.J.: Beyond regression: New tools for predictions and analysis in the behavioral sciences (Ph.D. thesis). Harvard University (1974)
70. Werbos, P.J.: The Roots of Backpropagation: From Ordered Derivatives to Neural Networks and Political Forecasting. Wiley, New York (1994)
71. Witten, I.H., Frank, E.: Data Mining: Practical Machine Learning Tools and Techniques. Morgan Kaufmann, Amsterdam (2005)
72. Xu, R., Wunsc, D.: Clustering. Wiley-IEEE Press, New York (2008)
73. Yager, R.R., Filev, D.: On the issue of defuzzification and selection based on a fuzzy set. Fuzzy Sets Syst. **55**, 255–271 (1993)
74. Yang, X.S., Cui, Z., Xiao, R., Gandomi, A.H., Karamanoglu, M.: Swarm Intelligence and Bio-Inspired Computation: Theory and Applications. Elsevier, Oxford (2013)
75. Yu, H., Wilamowski, B.M.: Levenberg–Marquardt training. The Industrial Electronics Handbook, Intelligent Systems **5**, (1963)
76. Yu, X.H., Chen, G.A., Cheng, S.X.: Acceleration of backpropagation learning using optimised learning rate and momentum. Electron. Lett. **29**, 1288–1290 (1993)
77. Yu, H.H., Chen, G.A., Cheng, S.X.: Dynamic learning rate optimization of the backpropagation algorithm. IEEE Trans. Neural Netw. **6**, 669–677 (1995)
78. Zadeh, L.A.: Fuzzy sets. Inf. Control **8**, 338–353 (1965)
79. Zheng, J., Li, M.: An efficient method for maintaining diversity in evolutionary multi-objective optimization, natural computation. In: Proceedings of the Fourth International Conference on Natural Computation (ICNC'08), vol. 6, pp. 462–467 (2008)
80. Zhenyu, Y., Lishan, K., Mckay, B., Penghui, F.: SEEA for multi-objective optimization: reinforcing elitist moea through multi-parent crossover, steady elimination and swarm hill climbing. In: Proceedings of the 4th Asia-Pasific Conference on Simulated Evolution and Learning, pp. 21–26 (2002)

Chapter 3
Introduction to Fuzzy System Interpretability

In general terms interpretability is a feature of a certain phenomenon, which facilitates its interpretation. This notion as such is imprecise because there are no commonly known solutions making it possible to unambiguously determine whether, for example, a fuzzy system is (and in particular its rules are) interpretable. The word "interpretability" could actually be substituted with a synonymous word of "readability". Of course, verification of interpretability also heavily depends on the initial knowledge and the basic cognitive structures, which are imagination and intelligence. This additionally complicates considerations from the field of interpretability making it still an open issue. Despite this fact, interpretability is important because it increases reliability and simplifies operation, allows us to refer the operation to the current state of knowledge, allows us to eliminate errors and inconsistencies in the operation, allows us to adapt to changing working conditions, etc.

In the field of modelling there is a division which refers to the possibility of interpretation of created models. It is a division of methods into three groups: "white-box", "gray-box" and "black-box". The modelling by white-box methods uses a phenomenological model, which is a mathematical description [8, 24, 31]. White-box models are readable, but often very simplified. Simplifying assumptions mainly concern idealization and linearization of characteristics, skipping the saturation phenomenon, skipping the friction phenomenon, etc. In the modelling by black-box methods the behavior of the object is modelled by cause-and-effect relationships [19, 32, 47]. Black-box models are accurate, but usually not interpretable. The most important from the point of view of interpretability are gray-box methods, in which we can obtain a satisfactory compromise between accuracy and interpretability. Gray-box models are often based on fuzzy systems. Their affiliation to gray-box methods results, among others, from a clear connection of antecedences and consequences of the rules and fundamental assumptions of Zadeh's fuzzy set theory [52].

When considering the issue of interpretability it should be emphasized that it is much more difficult for modelling tasks (regression) than the ones from the field of classification. This is due to the fact that in modelling the exact value of an output

K. Cpałka, *Design of Interpretable Fuzzy Systems*, Studies in Computational Intelligence 684, DOI 10.1007/978-3-319-52881-6_3

signal is important (see e.g. formula (2.18) or (2.19)), not rough indication of the class (see e.g. formula (2.20)). Thus, each constraint of the system structure used to enhance interpretability usually negatively affects the accuracy (and vice versa).

As already mentioned, there are no commonly known solutions for making it clear whether, for example, a fuzzy system, and in particular its rules, are interpretable. However, we can formulate a variety of interpretability criteria, regarded as a measure of interpretability. Assuming that a smaller value of these measures means better interpretability (e.g. lower complexity and better readability), we may use them, for example, in the process of automatic design of interpretable fuzzy systems. An example of such operation is described in Chap. 6.

This chapter describes the issue of accuracy, complexity and interpretability of fuzzy systems (Sect. 3.1). It contains also a review of solutions from the field of fuzzy systems interpretability (Sect. 3.2) and a reference to the attempts to systematize these solutions (Sect. 3.3). Moreover, it contains a review of basic (sample) interpretability criteria of fuzzy systems (Sect. 3.4). In the final part of the chapter a summary and a bibliography are presented (Sect. 3.5).

3.1 Fuzzy System Accuracy, Complexity and Interpretability

Fuzzy set theory provides a very good basis for designing interpretable fuzzy systems. This is due to the perception of reality by the man, who is inherently imprecise. The concept of a fuzzy set with clarity of rule-based notation fits perfectly into this imprecision. In the literature two approaches binding the issues of the fuzzy systems accuracy and their interpretability are considered (Fig. 3.1). They are the following approaches:

- Precise fuzzy modelling. The main purpose of this modelling is to obtain fuzzy models distinguished by a good accuracy (among others, in the sense of criteria considered in Sect. 2.3). An example of a system implementing this type of modelling is Takagi-Sugeno system (Takagi-Sugeno-Kang), in which generally consequences are functional dependencies binding input signals [7, 48].
- Linguistic fuzzy modelling. The main purpose of this modelling is to obtain fuzzy models distinguished by a good readability. The basis of this modelling are systems in which antecedents and consequences of rules are expressed by fuzzy sets that clearly define the values of input and output linguistic variables. An example of such system is Mamdani-type system described in Sect. 2.1.1 or logical-type system described in Sect. 2.1.2. It can be said that the systems implementing fuzzy linguistic modelling are inherently aimed at interpretability and they create better opportunities for action in this field.

When designing interpretable fuzzy systems we seek to ensure that they will be characterized by high accuracy, good readability of the knowledge accumulated in

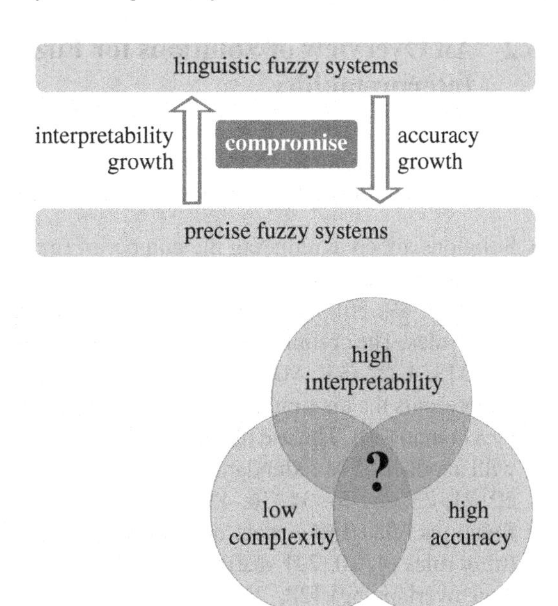

Fig. 3.1 A compromise (trade-off) between accuracy and interpretability, which is the basis for the division of fuzzy systems into precise and linguistic ones

Fig. 3.2 Summary of basic requirements imposed on interpretable fuzzy system. A question mark means a group of systems which meet all the requirements

them and low complexity (Fig. 3.2). However, development of such solution is not possible in practice for different reasons including the following two:

- In order to achieve high accuracy of modelling, we should use a complex structure of the fuzzy system (e.g. a larger rule base), which is not conducive to its readability.
- In order to obtain a good interpretability, we should use a possibly simple structure of the fuzzy system, which is not conducive to its accuracy.

Therefore, in practice the aim is to develop solutions enabling a satisfactory compromise between accuracy and interpretability of the system and the knowledge stored in it [1, 13, 15, 22, 27, 37, 40, 42, 51].

Considering the design aspects of interpretable fuzzy systems we should also pay attention to two types of interpretability:

- Interpretability aimed at the complexity of the system, in particular the complexity of its fuzzy rules base. In this case it is assumed that lower complexity of the fuzzy rule base means that the system interpretability is higher. This assumption is partially correct, but does not consider e.g. cohesion of fuzzy sets shape or overlapping fuzzy sets (much impeding their readability).
- Interpretability aimed at the readability of fuzzy rules base. Considering this type of interpretability we can analyze not only the consistency of the fuzzy sets shape or their coverage of consideration space, but also their adjustment to the training data, a way of fuzzy rules activation, etc.

3.2 An Overview of Solutions for Fuzzy Systems Interpretability

In the literature a number of solutions on the subject of interpretability of fuzzy systems can be found. Their authors have proposed e.g.:

- Solutions aimed at reducing the number of fuzzy rules [1, 3, 13, 16, 22, 26, 27], reducing the number of fuzzy sets [18] and aimed at reducing the number of system inputs [3, 49, 50]. Limitations were also related to the number of antecedents in fuzzy rules. The optimal number was most often set to Miller number, which equals 7 ± 2 [3, 4, 22, 33]. Miller number was designated in 1956 by George Miller and it represents the maximum pieces of information that can be directly distinguished by a human [30]. The use of restrictions in a system structure was often associated with a reduction of redundant elements and merging of similar ones [6, 9, 17, 18, 20, 21, 23, 25, 26, 34, 36, 43].
- Solutions related to correct notation of fuzzy rules [3, 26], correct activation of fuzzy rules [4, 10, 27], distinguishability and interdependence of fuzzy sets (e.g. their overlapping) [28, 29, 33] as well as solutions concerning such issues as complementarity, fitting in with data, etc. [5, 12, 13].
- Solutions related to fuzzy systems construction aimed at interpretability. In the papers [9, 35, 38, 45, 46] the use of additional weights of importance of the rules, antecedences, consequences and system inputs was proposed. In the paper [41] a dynamic structure of connections between fuzzy sets and rules was considered. It was proposed, among others, in order to reduce system complexity and to simplify the rule-based notation. In the papers [9, 11, 39] parametrized triangular norms were used (as precise aggregation operators) in order to increase accuracy, and in the paper [11] the authors reviewed parametrized triangular norms in terms of their suitability for the construction of fuzzy systems. In the paper [9] the authors used an extended (precise) defuzzification mechanism in which the number of discretization points does not have to be equal to the number of rules. This was suggested, among others, in order to increase accuracy of the system with a fixed number of rules and to provide opportunities to reduce the complexity of rules.

3.3 Attempts to Systematize Solutions for Fuzzy Systems Interpretability

The literature abounds in numerous attempts to systematize solutions for interpretability (e.g. [14, 42, 44]). The systematics presented in [14] deserves a special attention. Its authors have proposed division of solutions for interpretability into four groups - quadrants:

- The quadrant of solutions aimed at reducing complexity at fuzzy rules level. It takes into account, among others, the number of fuzzy rules, the number of antecedences in each rule and using Miller number.
- The quadrant of solutions aimed at reducing complexity at the fuzzy partitioning level. It takes into account, among others, the number of fuzzy sets associated with various inputs and outputs and the number of inputs.
- The quadrant concerning solutions aimed at increasing semantic readability at the fuzzy rules' level. It takes into account, among others, the consistency of the rules, activation level of rules and readability of rule-based notation.
- The quadrant concerning solutions aimed at increasing semantic readability at the fuzzy partitioning level. It takes into account, among others, a coverage degree of the input data by fuzzy sets, normalization of fuzzy sets, distinctness of fuzzy sets and fuzzy sets complementarity.

An interesting semantics has also been proposed in [2], which can complement the semantics proposed in [14]. In this type of semantics the solutions for interpretability were divided in terms of readability of the knowledge accumulated in the system and a different importance was symbolically assigned to the criteria. There are:

- Very important criteria for the complexity of fuzzy rules and notation readability.
- Important criteria for the semantics of rules and fuzzy sets, including the criteria for ordering fuzzy sets, semantic phrases used, sharing of fuzzy sets by a number of rules, etc.
- The least important criteria for the normalization of fuzzy sets, their complementarity, coverage of input data area, etc.

3.4 Basic Fuzzy System Interpretability Criteria

This section describes basic, exemplary interpretability criteria that can be used in the process of automatic design of interpretable fuzzy systems. They are, among others, the following criteria:

- Complexity criterion (Fig. 3.3). This criterion should be sufficient to assess complexity of a fuzzy system and take into account, among others, the number of fuzzy rules, input fuzzy sets, output fuzzy sets, inputs and discretization points.
- Criterion for assessing readability of fuzzy sets position (Fig. 3.4). The criterion should be sufficient to assess the correctness of input and output fuzzy sets distribution. Incorrect distribution of fuzzy sets may result from their overlapping and too much distance between them. The distribution of fuzzy sets can be evaluated, among others, by analyzing the intersection points of adjacent fuzzy sets. Therefore, the criterion may, for example, take into account two intersection points for each pair of adjacent fuzzy sets. In this case, taking into account the first intersection point would allow us to assess the distance between fuzzy sets, while taking

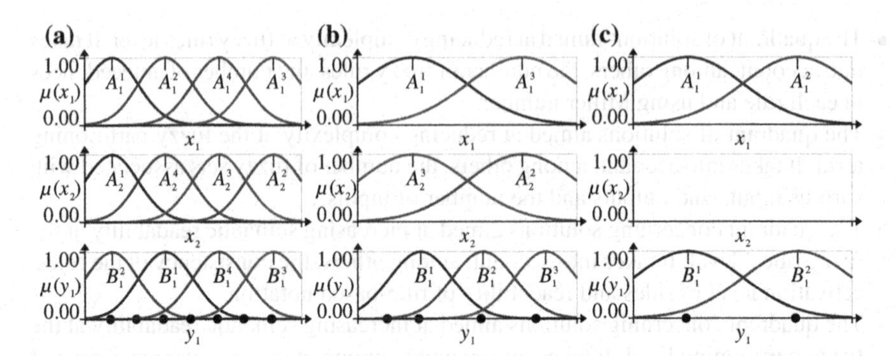

Fig. 3.3 Three examples illustrating the problem of the fuzzy system complexity: **a** unfavorable, **b** intermediate, **c** favorable (low complexity of the system, low criterion value). Discretization points are marked as *black circles*

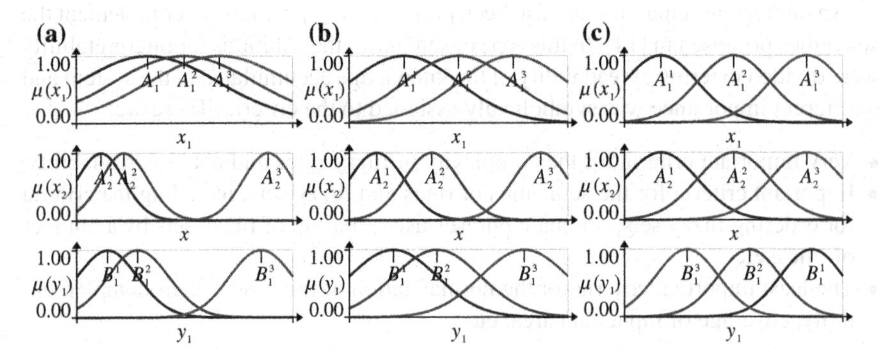

Fig. 3.4 Three examples illustrating the issue of fuzzy sets position readability: **a** unfavorable, **b** intermediate, **c** favorable (correct distribution of fuzzy sets, a low criterion value)

into account the second intersection point would allow us to evaluate the fuzzy sets overlapping.

- Criterion for assessing consistency of fuzzy sets width (Fig. 3.5). The criterion should be considered a cohesion measure of width of input and output fuzzy sets. This cohesion is important for the semantic readability of a fuzzy system rule base, because it facilitates understanding of a rule-based notation.
- The criterion for assessing uniformity of coverage of the data space by input fuzzy sets (Fig. 3.6). The criterion should be sufficient to assess the fit of the input fuzzy sets to the input data. Properly positioned input fuzzy sets associated with the input i are the ones for which the sum of membership determined for the signal of the input i is equal to 1. This assumption should be valid for all inputs of the system and should be evaluated in the context of the whole learning sequence.
- The criterion for assessing rules activity. The criterion should be sufficient to assess a way of the rule activation in the fuzzy system (Fig. 3.7). Correct activity of rules occurs when for each set from the learning sequence a single rule from the rule

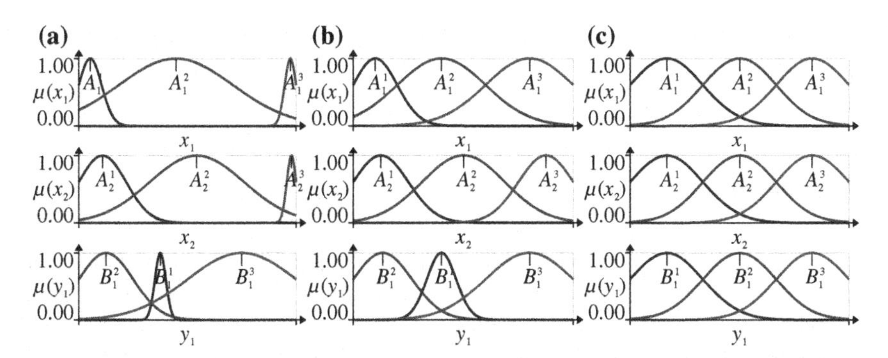

Fig. 3.5 Three examples illustrating the issue of fuzzy sets width coherence: **a** unfavorable, **b** intermediate, **c** favorable (minor differences between the widths of fuzzy sets, a low criterion value)

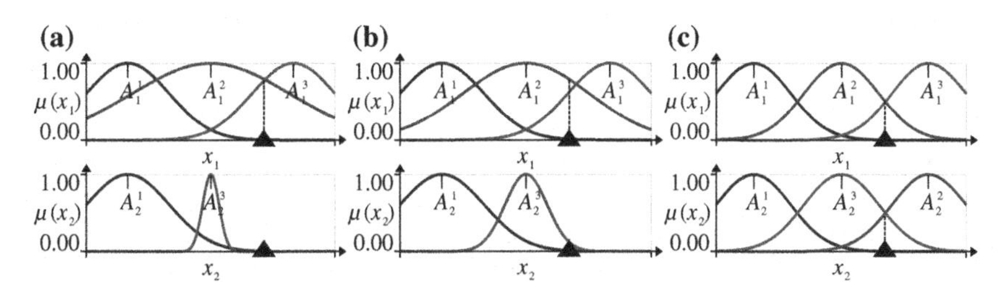

Fig. 3.6 Three examples illustrating the issue of uniformity of coverage of the data space by input fuzzy sets: **a** unfavorable, **b** intermediate, **c** favorable (good fit of fuzzy sets to the data, low criterion value). Location of signals from exemplary sample of learning sequence is marked as a triangle

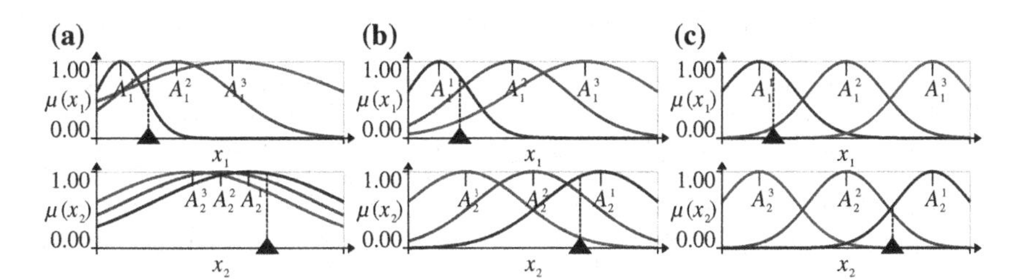

Fig. 3.7 Three examples illustrating the issue of readable rules activity: **a** unfavorable, **b** intermediate, **c** favorable (a small number of fuzzy rules activated at the same time, low criterion value). Location of signals from exemplary sample of learning sequence is marked as a triangle

base is activated with a minimal activation of the other rules. The activation level of rule k for the system (2.8) is expressed by the formula (2.6).

Assumptions of the interpretability measures mentioned above refer to the semantics considered in Sect. 3.3. They are supplemented by the criteria considered in Chap. 4. They refer to the specifics of the flexible type systems described in Chap. 4.

Then, some examples of the formulas defining the interpretability criteria are presented in Chap. 6.

3.5 Summary

In this chapter the relationship between accuracy, complexity and interpretability of fuzzy systems is described. Moreover, we present a review of current solutions from the field of fuzzy systems interpretability and described approaches which are well known in the literature and allow us to systematize developed solutions. In the final part of the chapter basic sample interpretability criteria of fuzzy systems are listed. It is worth noting that the solutions presented in the following chapters of this book fit well into the theme of interpretable fuzzy systems design, significantly expanding it in some areas.

References

1. Alcalá, R., Ducange, P., Herrera, F., Lazzerini, B., Marcelloni, F.: A multi-objective evolutionary approach to concurrently learn rule and data base soft linguistic fuzzy rule-based systems. IEEE Trans. Fuzzy Syst. **17**, 1106–1122 (2009)
2. Alonso, J.M.: Modeling Highly Interpretable Fuzzy Systems. Eur. Centre Soft Comput. (2010)
3. Alonso, J.M., Magdalena, L.: HILK++: an interpretability-guided fuzzy modeling methodology for learning readable and comprehensible fuzzy rule-based classifiers. Soft Comput. **15**, 1959–1980 (2011)
4. Alonso, J.M., Magdalena, L., Cordón, O.: Embedding HILK in a three-objective evolutionary algorithm with the aim of modeling highly interpretable fuzzy rule-based classifiers. In: Proceedings of the 4th International Workshop on Genetic and Evolving Fuzzy Systems (GEFS2010), pp. 15–20 (2010)
5. Botta, A., Lazzerini, B., Marcelloni, F., Stefanescu, D.C.: Context adaptation of fuzzy systems through a multi-objective evolutionary approach based on a novel interpretability index. Soft Comput. **13**, 437–449 (2009)
6. Casillas, J., Cordon, O., Herrera, F., Magdalena, L. (eds.): Interpretability Issues in Fuzzy Modeling. Springer, New York (2003)
7. Chadli, M., Borne, P.: Multiple Models Approach in Automation: Takagi-Sugeno Fuzzy Systems. Wiley, New Jersey (2012)
8. Chen, K., Bouscayrol, A., Berthon, A., Delarue, P., Hissel, D., Trigui, R.: Global modeling of different vehicles. IEEE Veh. Technol. Mag. **4**, 80–89 (2009)
9. Cpałka, K.: A new method for design and reduction of neuro-fuzzy classification systems. IEEE Trans. Neural Netw. **20**, 701–714 (2009)
10. Espinosa, J., Vandewalle, J.: Constructing fuzzy models with linguistic integrity from numerical data-AFRELI algorithm. IEEE Trans. Fuzzy Syst. **8**, 591–600 (2000)
11. Farahbod, F., Eftekhari, M.: Comparsion of different T-norm operators in classification problems. Int. J. Fuzzy Logic Syst. **2**, 33–41 (2012)
12. Fazendeiro, P., De Oliveira, J.V., Pedrycz, W.: A multiobjective design of a patient and anaesthetist-friendly neuromuscular blockade controller. IEEE Trans. Biomed. Eng. **54**, 1667–1678 (2007)

13. Gacto, M.J., Alcalá, R., Herrera, F.: Integration of an index to preserve the semantic interpretability in the multi-objective evolutionary rule selection and tuning of linguistic fuzzy systems. IEEE Trans. Fuzzy Syst. **18**, 515–531 (2010)
14. Gacto, M.J., Alcalá, R., Herrera, F.: Interpretability of linguistic fuzzy rule-based systems: an overview of interpretability measures. Inf. Sci. **181**, 4340–4360 (2011)
15. Gorzałczany, M., Rudziński, F.: A multi-objective genetic optimization for fast, fuzzy rule-based credit classification with balanced accuracy and interpretability. Appl. Soft Comput. **40**, 206–220 (2016)
16. Gorzałczany, M.B., Rudziński, F.: Accuracy versus interpretability of fuzzy rule-based classifiers: an evolutionary approach. In: Proceedings of the International Conference on Swarm and Evolutionary Computation (SIDE'12), pp. 222–230 (2012)
17. Guillaume, S.: Designing fuzzy inference systems from data: an interpretability oriented review. IEEE Trans. Fuzzy Syst. **9**, 426–443 (2001)
18. Guillaume, S., Charnomordic, B.: Generating an interpretable family of fuzzy partitions from data. IEEE Trans. Fuzzy Syst. **12**, 324–335 (2004)
19. Ibrahim, S.S., Bamatraf, M.,A.: Interpretation trained neural networks based on genetic algorithms. Int. J. Artif. Intell. Appl. (IJAIA) **4**, 13–22 (2013)
20. Icke, I., Rosenberg, A.: Multi-objective Genetic Programming for Visual Analytics. Lecture Notes in Computer Science, vol. 6621, pp. 322–334. Springer, Heidelberg (2011)
21. Ishibuchi, H., Nakashima, T., Murata, T.: Performance evaluation of fuzzy classifier systems for multidimensional pattern classification problems. IEEE Trans. Syst. Man Cybern. Part B **29**, 601–618 (1999)
22. Ishibuchi, H., Nojima, Y.: Analysis of interpretability-accuracy tradeoff of fuzzy systems by multiobjective fuzzy genetics-based machine learning. Int. J. Approx. Reason. **44**, 4–31 (2007)
23. Jin, Y.: Fuzzy modeling of high dimensional systems: complexity reduction and interpretability improvement. IEEE Trans. Fuzzy Syst. **8**, 212–221 (2000)
24. Kaczorek, T.: A modified state variable diagram method for determination of positive realizations of linear continous-time systems with delays. Int. J. Appl. Math. Comput. Sci. **22**, 897–905 (2012)
25. Kenesei, T., Abonyi, J.: Interpretable support vector machines in regression and classification - application in process engineering. Hung. J. Ind. Chem. **35**, 101–108 (2007)
26. Liu, F., Quek, C., Ng, G.S.: A novel generic hebbian ordering-based fuzzy rule base reduction approach to Mamdani neuro-fuzzy system. Neural Comput. **19**, 1656–1680 (2007)
27. Marquez, A.A., Marquez, F.A., Peregrin, A.: A multi-objective evolutionary algorithm with an interpretability improvement mechanism for linguistic fuzzy systems with adaptive defuzzification. In: Proceedings of the IEEE International Conference on Fuzzy Systems (FUZZ2010), pp. 1-7 (2010)
28. Mencar, C., Castellano, G., Fanelli, A.M.: On the role of interpretability in fuzzy data mining. Int. J. Uncertain. Fuzziness Knowl. Based Syst. **15**, 521–537 (2007)
29. Mencar, C., Castiello, C., Cannone, R., Fanelli, A.M.: Interpretability assessment of fuzzy knowledge bases: a cointension based approach. Int. J. Appro. Reason. **52**, 501–518 (2011)
30. Miller, G.A.: The magical number seven, plus or minus two: some limits on our capacity for processing information. Psychol. Rev. **63**, 81–97 (1956)
31. Musa, A.A.H., Muawia, M.A.: Analysis of the dc motor speed control using state variable transition matrix. Int. J. Sci. Res. (IJSR) **3**, 2758–2763 (2014)
32. Patan, K., Korbicz, J.: Nonlinear model predictive control of a boiler unit: a fault tolerant control study. Int. J. Appl. Math. Comput. Sci. **22**, 225–237 (2012)
33. Pulkkinen, P., Koivisto, H.: A dynamically constrained multiobjective genetic fuzzy system for regression problems. IEEE Trans. Fuzzy Syst. **18**, 161–177 (2010)
34. Riid, A., Rustern, E.: Interpretability improvement of fuzzy systems: reducing the number of unique singletons in zeroth order Takagi-Sugeno systems. In: Proceedings of the IEEE International Conference on Fuzzy Systems, pp. 1–6 (2010)
35. Riid, A., Rustern, E.: Interpretability, Interpolation and Rule Weights in Linguistic Fuzzy Modeling. Lecture Notes in Computer Science, vol. 6857, pp. 91–98. Springer, Heidelberg (2011)

36. Roubos, H., Setnes, M.: Compact and transparent fuzzy models and classifiers through iterative complexity reduction. IEEE Trans. Fuzzy Syst. **9**, 516–524 (2001)
37. Rudziński, F.: A multi-objective genetic optimization of interpretability-oriented fuzzy rule-based classifiers. Appl. Soft Comput. **38**, 118–133 (2016)
38. Rutkowski, L.: Computational Intelligence. Springer, Heidelberg (2010)
39. Rutkowski, L., Cpałka, K.: Flexible neuro fuzzy systems. IEEE Trans. Neural Netw. **14**, 554–574 (2003)
40. Sánchez, G., Jiménez, F., Sánchez, J.M., Alcaraz, J.M.: A Multi-objective Neuro-evolutionary Algorithm to Obtain Interpretable Fuzzy Models. Lecture Notes in Computer Science, vol. 5988, pp. 51–60. Springer, Heidelberg (2010)
41. Scherer, R.: Neuro-fuzzy Systems with Relation Matrix. Lecture Notes in Computer Science, vol. 6113, pp. 210–215. Springer, Heidelberg (2010)
42. Shukla, P.K., Tripathi, S.P.: A review on the interpretability-accuracy trade-off in evolutionary multi-objective fuzzy systems (EMOFS). Information **3**, 256–277 (2012)
43. Shukla, P.K., Tripathi, S.P.: Handling high dimensionality and interpretability-accuracy trade-off issues in evolutionary multiobjective fuzzy classifiers. Int. J. Sci. Eng. Res. **5**, 665–671 (2014)
44. Shukla, P.K., Tripathi, S.P.: A new approach for tuning interval type-2 fuzzy knowledge bases using genetic algorithms. J. Uncertain. Anal. Appl. **2**, 1–15 (2014)
45. Simiński, K.: Rule weights in a neuro-fuzzy system with a hierarchical domain partition. Appl. Math. Comput. Sci. **20**, 337–347 (2010)
46. Singh, L., Kumar, S., Paul, S.: Automatic simultaneous architecture and parameter search in fuzzy neural network learning using novel variable length crossover differential evolution. In: Proceedings of the IEEE International Conference on Fuzzy Systems, pp. 1795–1802 (2008)
47. Tadeusiewicz, R.: Place and role of intelligent systems in computer science. Comput. Method. Mater. Sci. **10**, 193–206 (2010)
48. Takagi, T., Sugeno, M.: Fuzzy identification of systems and its application to modeling and control. IEEE Trans. Syst. Man Cybern. **15**, 116–132 (1985)
49. Tikk, D., Gedeon, T., Wong, K.: A feature ranking algorithm for fuzzy modeling problems. In: Casillas, J., Cordón, O., Herrera, F., Magdalena, L. (eds.) Interpretability Issues in Fuzzy Modeling, pp. 176–192. Springer-Verlag, Heidelberg (2003)
50. Vanhoucke, V., Silipo, R.: Interpretability in multidimensional classification. In: Casillas, J., Cordón, O., Herrera, F., Magdalena, L. (eds.) Interpretability Issues in Fuzzy Modeling, pp. 193–217. Springer-Verlag, Heidelberg (2003)
51. Wang, H., Kwong, S., Jin, Y., Wei, W., Man, K.F.: Multi-objective hierarchical genetic algorithm for interpretable fuzzy rule-based knowledge extraction. Fuzzy Sets Syst. **149**, 149–186 (2005)
52. Zadeh, L.A.: Fuzzy sets Inf. Control **8**, 338–353 (1965)

Chapter 4
Improving Fuzzy Systems Interpretability by Appropriate Selection of Their Structure

Affecting interpretability of fuzzy systems by appropriate selection of their structure is an issue which could prove to be crucial in designing interpretable fuzzy systems. The reason for that is obvious. When designing a system we seek to achieve a required accuracy. If the accuracy happens to be less than satisfactory, then a typical action which follows is the development of a fuzzy system rule base. However, it does reduce the base interpretability. The solution proposed in this chapter, which is used to increase the accuracy of a system without expanding its rule base, is to increase the precision of aggregation, inference, and defuzzification operators. This leads to the development of a new class of fuzzy systems, called flexible-type systems [14–17, 41–43]. In this book operators with an increased precision will be referred to as precise operators. However, it does not mean to imply that other operators should be seen as imprecise.

Flexibilisation of fuzzy systems (for example the systems described in Chap. 2) may include a variety of measures aimed at increasing their interpretability by increasing operation precision. Additionally, this has many other advantages, e.g.:

- The possibility to increase the versatility of fuzzy rules description. Owing it to the specific aspects featured by flexibility, such description can take into account, among others, weights of importance of input fuzzy sets, output fuzzy sets and also the whole rules. This increases experts' possibilities within the scope of the description of a problem (formulation of fuzzy rules). It also allows us to introduce a hierarchy of rules in the rule base, which is impossible to achieve in the classic notation of rules in the form (2.2). This hierarchy can be generated automatically during a learning process.
- The possibility to simplify the system structure selection procedure. Thanks to the advantages offered by flexibility, for example, parameterized triangular norms can be used in the system, which minimizes the risk of improper selection of non-flexible norms. These norms can be automatically adjusted during a learning process.

K. Cpałka, *Design of Interpretable Fuzzy Systems*, Studies in Computational Intelligence 684, DOI 10.1007/978-3-319-52881-6_4

- The possibility to automatically reduce a rule base without affecting the defuzzification process (which usually results in a loss of accuracy). It can result e.g. from the use of a modified defuzzification formula [15, 16], in which it is not necessary to preserve equality between the number of rules and the number of discretization points of the defuzzification operator (e.g. the operator of the form (2.7)). Among its other functions, this equality blocked the possibility to reduce redundant system rules.
- The possibility to automatically select the inference type for the considered problem. Appropriate operators facilitate its automatic adjustment during a learning process.

The author of this book has already contributed to the considerations on fuzzy systems flexibilisation in the existing literature on the subject (see e.g. [14–17, 41–43]). However, so far it has not been treated as a mechanism aimed at increasing interpretability of fuzzy systems.

This chapter discusses precise aggregation, inference (Sect. 4.1) and defuzzification (Sect. 4.2) operators. Moreover, it describes fuzzy systems resulting from the use of these operators (Sect. 4.3) and interpretability criteria dedicated to them (Sect. 4.4). The final part of the chapter contains examples of simulation results (Sect. 4.5), a summary (Sect. 4.6) and bibliography.

4.1 Precise Aggregation and Inference Operators

Literature presents a number of different operators which can be used as precise aggregation or inference operators [14, 15, 41–43]. This section presents characteristics of the chosen operators which are used to construct flexible-type fuzzy systems.

4.1.1 Parameterized Triangular Norms

In practical applications of the fuzzy set theory, we can use parameterized variants of triangular norms [19, 26, 29, 36, 39, 40, 42, 46, 48, 49] which meet all the conditions required by the triangular norms [4, 27, 28] (i.e. monotonicity, commutativity, associativity and boundary). They include the triangular norms of Dombi, Hamacher, Yager, Frank, Weber, Dubois and Prady, Schweizer, Mizumoto, etc. The selected ones are presented in the Tables 4.1 and 4.2.

One of the characteristic features of the parameterized triangular norms is that the shape of their hyperplanes may be modified as a result of changing the value of a certain parameter p within the range from the min/max-type norms to the drastic type norms (Table 2.1). Selection of the parameter p value can be performed automatically during a learning process. Allowable range of changes of the parameter p value

Table 4.1 A list of parameterized triangular norms (part I)

Name	t-norm and t-conorm	p
Dombi	$\overleftrightarrow{T}\{a_1, a_2; p\} = \frac{1}{1+\left(\left(\frac{1}{a_1}-1\right)^p+\left(\frac{1}{a_2}-1\right)^p\right)^{\frac{1}{p}}}$	$p > 0$
	$\overleftrightarrow{S}\{a_1, a_2; p\} = \frac{1}{1+\left(\left(\frac{1}{a_1}-1\right)^{-p}+\left(\frac{1}{a_2}-1\right)^{-p}\right)^{-\frac{1}{p}}}$	
Hamacher	$\overleftrightarrow{T}\{a_1, a_2; p\} = \frac{a_1 \cdot a_2}{p+(1-p)\cdot(a_1+a_2-a_1 \cdot a_2)}$	$p \geq 0$
	$\overleftrightarrow{S}\{a_1, a_2; p\} = \frac{a_1+a_2-a_1 \cdot a_2-(1-p)\cdot a_1 \cdot a_2}{1-(1-p)\cdot a_1 \cdot a_2}$	
Yager	$\overleftrightarrow{T}\{a_1, a_2; p\} = 1 - \min\left\{1, \sqrt[p]{(1-a_1)^p + (1-a_2)^p}\right\}$	$p > 0$
	$\overleftrightarrow{S}\{a_1, a_2; p\} = \min\left\{1, \sqrt[p]{a_1{}^p + a_2{}^p}\right\}$	
Frank	$\overleftrightarrow{T}\{a_1, a_2; p\} = \log_p\left(1 + \frac{(p^{a_1}-1)\cdot(p^{a_2}-1)}{p-1}\right)$	$p > 0$, $p \neq 1$
	$\overleftrightarrow{S}\{a_1, a_2; p\} = 1 - \log_p\left(1 + \frac{(p^{1-a_1}-1)\cdot(p^{1-a_2}-1)}{p-1}\right)$	
Weber I	$\overleftrightarrow{T}\{a_1, a_2; p\} = \max\left\{0, \frac{a_1+a_2-1+p \cdot a_1 \cdot a_2}{1+p}\right\}$	$p > -1$
	$\overleftrightarrow{S}\{a_1, a_2; p\} = \min\left\{1, \frac{(1+p)\cdot(a_1+a_2)-p \cdot a_1 \cdot a_2}{1+p}\right\}$	
Weber II	$\overleftrightarrow{T}\{a_1, a_2; p\} = \max\left\{\begin{array}{c} 0, (1+p)\cdot a_1 + (1+p)\cdot a_2 + \\ -p \cdot a_1 \cdot a_2 - (1+p) \end{array}\right\}$	$p > -1$
	$\overleftrightarrow{S}\{a_1, a_2; p\} = \min\{1, a_1 + a_2 + p \cdot a_1 \cdot a_2\}$	
Dubois and Prade	$\overleftrightarrow{T}\{a_1, a_2; p\} = \frac{a_1 \cdot a_2}{\max\{a_1, a_2, p\}}$	$0 \leq p \leq 1$
	$\overleftrightarrow{S}\{a_1, a_2; p\} = 1 - \frac{(1-a_1)\cdot(1-a_2)}{\max\{(1-a_1),(1-a_2),p\}}$	
Schweizer I	$\overleftrightarrow{T}\{a_1, a_2; p\} = \sqrt[p]{\max\{0, a_1{}^p + a_2{}^p - 1\}}$	$p > 0$
	$\overleftrightarrow{S}\{a_1, a_2; p\} = 1 - \sqrt[p]{\max\{0, (1-a_1)^p + (1-a_2)^p - 1\}}$	
Schweizer II	$\overleftrightarrow{T}\{a_1, a_2; p\} = \frac{1}{\sqrt[p]{\frac{1}{a_1{}^p}+\frac{1}{a_1{}^p}-1}}$	$p > 0$
	$\overleftrightarrow{S}\{a_1, a_2; p\} = 1 - \frac{1}{\sqrt[p]{\frac{1}{(1-a_1)^p}+\frac{1}{(1-a_1)^p}-1}}$	

Table 4.2 A list of parameterized triangular norms (part II)

Name	t-norm and t-conorm	p
Schweizer III	$\overset{\leftrightarrow}{T}\{a_1, a_2; p\} = 1 - \sqrt[p]{(1-a_1)^p + (1-a_2)^p - (1-a_1)^p \cdot (1-a_2)^p}$	$p > 0$
	$\overset{\leftrightarrow}{S}\{a_1, a_2; p\} = \sqrt[p]{a_1{}^p + a_2{}^p - a_1{}^p \cdot a_2{}^p}$	
Mizumoto IV	$\overset{\leftrightarrow}{T}\{a_1, a_2; p\} = \begin{cases} \frac{1}{p} \cdot \left(\frac{1}{\max\left(1, \frac{1}{1+p \cdot a_1} + \frac{1}{1+p \cdot a_2} - \frac{1}{1+p}\right)} - 1 \right) \\ \text{if } -1 < p < 0 \\ \frac{1}{p} \cdot \left(\frac{1}{\min\left(1, \frac{1}{1+p \cdot a_1} + \frac{1}{1+p \cdot a_2} - \frac{1}{1+p}\right)} - 1 \right) \\ \text{if } 0 < p \end{cases}$	$p > -1$
	$\overset{\leftrightarrow}{S}\{a_1, a_2; p\} = \begin{cases} 1 - \frac{1}{p} \cdot \left(\frac{1}{\max\left(1, \frac{1}{1+p \cdot (1-a_1)} + \frac{1}{1+p \cdot (1-a_2)} - \frac{1}{1+p}\right)} - 1 \right) \\ \text{if } -1 < p < 0 \\ 1 - \frac{1}{p} \cdot \left(\frac{1}{\min\left(1, \frac{1}{1+p \cdot (1-a_1)} + \frac{1}{1+p \cdot (1-a_2)} - \frac{1}{1+p}\right)} - 1 \right) \\ \text{if } 0 < p \end{cases}$	
Mizumoto V	$\overset{\leftrightarrow}{T}\{a_1, a_2; p\} = 1 - \log_p \left(\min\left\{p, p^{1-a_1} + p^{1-a_2} - 1\right\}\right)$	$p > 1$
	$\overset{\leftrightarrow}{S}\{a_1, a_2; p\} = \log_p (\min\{p, p^{a_1} + p^{a_2} - 1\})$	
Mizumoto VI	$\overset{\leftrightarrow}{T}\{a_1, a_2; p\} = 1 - \frac{1}{p} \cdot \ln\left(\min\left\{e^p, e^{p \cdot (1-a_1)} + e^{p \cdot (1-a_2)} - 1\right\}\right)$	$p > 0$
	$\overset{\leftrightarrow}{S}\{a_1, a_2; p\} = \frac{1}{p} \cdot \ln (\min\{e^p, e^{p \cdot a_1} + e^{p \cdot a_2} - 1\})$	
Mizumoto VII	$\overset{\leftrightarrow}{T}\{a_1, a_2; p\} = \sqrt[p]{1 - \ln\left(\min\left\{e, e^{1-a_1{}^p} + e^{1-a_2{}^p} - 1\right\}\right)}$	$p > 0$
	$\overset{\leftrightarrow}{S}\{a_1, a_2; p\} = 1 - \sqrt[p]{1 - \ln\left(\min\left\{e, e^{1-(1-a_1)^p} + e^{1-(1-a_2)^p} - 1\right\}\right)}$	
Mizumoto VII	$\overset{\leftrightarrow}{T}\{a_1, a_2; p\} = \frac{1}{\log_p\left(p^{\frac{1}{a_1}} + p^{\frac{1}{a_2}} - p\right)}$	$p > 1$
	$\overset{\leftrightarrow}{S}\{a_1, a_2; p\} = 1 - \frac{1}{\log_p\left(p^{\frac{1}{1-a_1}} + p^{\frac{1}{1-a_2}} - p\right)}$	
Mizumoto IX	$\overset{\leftrightarrow}{T}\{a_1, a_2; p\} = \frac{1}{\frac{1}{p} \cdot \ln\left(e^{\frac{p}{a_1}} + e^{\frac{p}{a_2}} - e^p\right)}$	$p > 0$
	$\overset{\leftrightarrow}{S}\{a_1, a_2; p\} = 1 - \frac{1}{\frac{1}{p} \cdot \ln\left(e^{\frac{p}{1-a_1}} + e^{\frac{p}{1-a_2}} - e^p\right)}$	
Mizumoto X	$\overset{\leftrightarrow}{T}\{a_1, a_2; p\} = \frac{1}{\sqrt[p]{\ln\left(e^{\frac{1}{a_1{}^p}} + e^{\frac{1}{a_2{}^p}} - e\right)}}$	$p > 0$
	$\overset{\leftrightarrow}{S}\{a_1, a_2; p\} = 1 - \frac{1}{\sqrt[p]{\ln\left(e^{\frac{1}{(1-a_1)^p}} + e^{\frac{1}{(1-a_2)^p}} - e\right)}}$	

depends on the type of the parameterized triangular norms (Tables 4.1 and 4.2). They are denoted as $\overleftrightarrow{T}\{\mathbf{a};\, p\}$ and $\overleftrightarrow{S}\{\mathbf{a};\, p\}$.

An example of the parameterized triangular norms are Dombi triangular norms. Dombi T-norm can be expressed as follows:

$$\begin{cases} \overleftrightarrow{T}_D\{\mathbf{a};\, p\} = \begin{cases} T_d\{\mathbf{a}\} & \text{for} \quad p=0 \\ \left(1+\left(\sum_{i=1}^{n}\left(\frac{1-a_i}{a_i}\right)^p\right)^{\frac{1}{p}}\right)^{-1} & \text{for } p\in(0,\infty) \\ T_m\{\mathbf{a}\} & \text{for} \quad p=\infty \end{cases} \\ \overleftrightarrow{S}_D\{\mathbf{a};\, p\} = \begin{cases} S_d\{\mathbf{a}\} & \text{for} \quad p=0 \\ 1-\left(1+\left(\sum_{i=1}^{n}\left(\frac{a_i}{1-a_i}\right)^p\right)^{\frac{1}{p}}\right)^{-1} & \text{for } p\in(0,\infty) \\ S_m\{\mathbf{a}\} & \text{for} \quad p=\infty. \end{cases} \end{cases} \tag{4.1}$$

Another example of parameterized triangular norms are Yager norms, which can be expressed as follows:

$$\begin{cases} \overleftrightarrow{T}_Y\{\mathbf{a};\, p\} = \begin{cases} T_d\{\mathbf{a}\} & \text{for} \quad p=0 \\ \max\left\{0,\, 1-\left(\sum_{i=1}^{n}(1-a_i)^p\right)^{\frac{1}{p}}\right\} & \text{for } p\in(0,\infty) \\ T_m\{\mathbf{a}\} & \text{for} \quad p=\infty \end{cases} \\ \overleftrightarrow{S}_Y\{\mathbf{a};\, p\} = \begin{cases} S_d\{\mathbf{a}\} & \text{for} \quad p=0 \\ \min\left\{1,\, \left(\sum_{i=1}^{n}(a_i)^p\right)^{\frac{1}{p}}\right\} & \text{for } p\in(0,\infty) \\ S_m\{\mathbf{a}\} & \text{for} \quad p=\infty. \end{cases} \end{cases} \tag{4.2}$$

It is worth noting that the parameterized t-norms for $n=2$ can be inference operators, for example, in Mamdani-type systems considered in Sect. 2.1.1. Then, by combining the idea of a parameterized t-conorm with the idea of s-implication we obtain parameterized s-implications, which can work as inference operators, for example, in logical-type systems considered in Sect. 2.1.2.

Properties of the parameterized triangular norms as well as those of inference operators implemented on their basis are especially valuable in the fuzzy system design phase. They can limit the negative impact of poor selection of the type of non-parameterized triangular norms on system accuracy. This is related to the existence of many varieties of non-parameterized triangular norms, which can be approximated by parameterized norms for specific values of the parameter p. Since this parameter can be selected during a learning process automatically, a parameterized norm can be replaced by a suitable non-parameterized counterpart on completion of the learning process. In the literature some attempts to compare different triangular norms in the context of their impact on the accuracy of fuzzy systems based on them can be found [20].

In designing interpretable fuzzy systems the use of parameterized triangular norms can contribute to increasing the precision of aggregation and inference operators, ensuring, for example, the required accuracy of the system with a smaller fuzzy rules base.

4.1.2 Triangular Norms with Weights of Arguments

In practical applications of the fuzzy set theory, the structure of fuzzy systems is often extended by weights of importance [1–3, 5, 7–13, 15, 16, 21–25, 30, 32, 37–40, 42, 45, 47, 50, 51]. It is obvious that adding new parameters to a system (including weights) usually has a beneficial effect on its accuracy. However, if these parameters do not have their interpretation, then such fuzzy system works like the methods from the black-box group [37, 38]. The triangular norms with weights of arguments proposed in this section are free from these disadvantages: they have their specified interpretation in the rule base and aggregation operators. Moreover, they can have dedicated interpretability criteria formulated for them (they are discussed in Sect. 4.4).

The notations $T^* \{\mathbf{a}; \mathbf{w}\}$ and $S^* \{\mathbf{a}; \mathbf{w}\}$ will be used to denote triangular norms with weights of arguments. Their description also includes the fact that their arguments $\mathbf{a} = [a_1, \ldots, a_n]$ ($a_i \in [0, 1]$) are related to weights of importance $\mathbf{w} = [w_1, \ldots, w_n]$ ($w_i \in [0, 1]$). The t-norm with weights of arguments $T^* \{\cdot\}$ is defined as follows:

$$T^* \{\mathbf{a}; \mathbf{w}\} = \mathop{T}_{i=1}^{n} \{S \{a_i, \operatorname{neg}(w_i)\}\}, \tag{4.3}$$

where $T \{\cdot\}$ is a base t-norm (on the basis of which a t-norm with weights has been designed), $S \{\cdot\}$ is any t-conorm (it does not have to be dual to the base t-norm used [4, 27, 28]), and $\operatorname{neg}(\cdot)$ is a negation operator. The operator of the form (4.3) satisfies monotonicity, commutativity and associativity for the t-norm [4, 27, 28] and the following boundary condition:

$$T^* \{a_1, 1; w_1, w_2\} = S \{a_1, \operatorname{neg}(w_1)\}. \tag{4.4}$$

The condition of the form (4.4) can be regarded as a boundary condition for the t-norm (i.e. $T \{1, a_1\} = a_1$) when $w_1 = 1$. The other remarks on the t-norm with weights of arguments can be summarized as follows:

- If weights of all the arguments of the operator (4.3) are equal to 1 ($\mathbf{w} = \mathbf{1}$), then it is reduced to a non-flexible t-norm ($T^* \{\mathbf{a}; \mathbf{1}\} = T \{\mathbf{a}\}$).
- If the weight of any argument of the operator (4.3) is equal to 0, then under the boundary condition of the t-conorm ($S \{1, a\} = 1$) and the boundary condition of the t-norm ($T \{1, a\} = a$), the argument is reduced (e.g. $T^* \{a_1, a_2; w_1, 0\} = a_1$).

The t-conorm with weights of arguments $S^* \{\cdot\}$ can be denoted analogously. It is defined as follows:

$$S^* \{\mathbf{a}; \mathbf{w}\} = \mathop{S}_{i=1}^{n} \{T \{a_i, w_i\}\}, \tag{4.5}$$

where $S \{\cdot\}$ is a base t-conorm (on the basis of which a t-conorm with weights has been designed) and $T \{\cdot\}$ is any t-norm (it does not have to be dual to the base t-conorm used).

The operator of the form (4.5) satisfies monotonicity, commutativity and associativity for the t-conorm [4, 27, 28] and the following boundary condition:

$$S^* \{a_1, 0; w_1, w_2\} = T \{a_1, w_1\}. \tag{4.6}$$

The condition of the form (4.6) can be regarded as a boundary condition for the t-conorm (i.e. $S \{0, a_1\} = a_1$) when $w_1 = 1$. The other remarks on the t-conorm with weights of arguments can be summarized as follows:

- If weights of all the arguments of the operator (4.5) are equal to 1 ($\mathbf{w} = \mathbf{1}$), then it is reduced to non-flexible t-conorm ($S^* \{\mathbf{a}; \mathbf{1}\} = S \{\mathbf{a}\}$).
- If the weight of any argument of the operator (4.5) is equal to 0, then the argument is reduced (e.g. $S^* \{a_1, a_2; w_1, 0\} = a_1$).

An example of triangular norms with weights of arguments are min/max norms with weights of arguments constructed on the basis of Zadeh negation (2.14) [44, 52]. They are described by the following relations:

$$\begin{cases} T_m{}^* \{\mathbf{a}; \mathbf{w}\} = \min\limits_{i=1,\dots,n} \{\max \{a_i, 1 - w_i\}\} \\ S_m{}^* \{\mathbf{a}; \mathbf{w}\} = \max\limits_{i=1,\dots,n} \{\min \{a_i, w_i\}\}. \end{cases} \tag{4.7}$$

Another example of triangular norms with weights of arguments are algebraic norms with weights of arguments constructed on the basis of Zadeh negation (2.14). They are described by the following relations:

$$\begin{cases} T_a{}^* \{\mathbf{a}; \mathbf{w}\} = \prod\limits_{i=1}^{n} (1 + (a_i - 1) \cdot w_i) \\ S_a{}^* \{\mathbf{a}; \mathbf{w}\} = 1 - \prod\limits_{i=1}^{n} (1 - a_i \cdot w_i). \end{cases} \tag{4.8}$$

It is worth noting that t-norms with weights of arguments for $n = 2$ can work as inference operators, for example, in Mamdani-type systems discussed in Sect. 2.1.1. Then, combining the idea of a t-conorm with weights of arguments with the idea of an s-implication (similar to the case of parameterized triangular norms) we obtain s-implications with weights of arguments which allow us to account for the importance of output fuzzy sets in an inference process. They have the following form:

$$I^*(a_1, a_2; w_1, w_2) = S^*\{\text{neg}(a_1), a_2; w_1, w_2\}. \tag{4.9}$$

An example of an s-implication with weights of arguments constructed on the basis of Zadeh negation (2.14) and a t-conorm of max type is the Kleene-Dienes s-implication with weights of arguments, which is defined as follows:

$$I_m^*(a_1, a_2; w_1, w_2) = \max\{\min\{1 - a_1, w_1\}, \min\{a_2, w_2\}\}. \tag{4.10}$$

Another example of an s-implication with weights of arguments constructed on the basis of Zadeh negation (2.14) and an algebraic t-conorm is Reichenbach s-implication with weights of arguments, which is defined as follows:

$$I_a^*(a_1, a_2; w_1, w_2) = 1 - (1 - (1 - a_1) \cdot w_1) \cdot (1 - a_2 \cdot w_2). \tag{4.11}$$

Operators of the form (4.9) can be inference operators, for example, in logical-type systems presented in Sect. 2.1.2.

The properties of triangular norms with weights of arguments are valuable from the point of view of the operation of fuzzy systems. First of all, they allow us to determine the importance of fuzzy rules (e.g. in the context of formula (2.11) or (2.16)), output fuzzy sets (e.g. in the context of formula (2.9) or (2.13)) and input fuzzy sets (e.g. in the context of formula (2.6)). Moreover, they create interesting opportunities for reducing the fuzzy rule base.

4.1.3 Switchable Triangular Norms

In practical applications of the fuzzy set theory there might occur a situation in which it is difficult to choose not only the type of the aggregation operators used but also the type of inference (e.g. Mamdani-type inference or logical-type inference). In such case we can use norm-switchable triangular norms [39, 42]. These norms may vary depending on the value of their parameter between the t-norm and t-conorm. An example of such function is an h-function constructed on the basis of the duality of triangular norms described by two alternative (for dual norms [4, 27, 28]) formulas:

$$\begin{aligned}\tilde{H}(\mathbf{a}; \nu) &= \widetilde{\text{neg}}\left(\overset{n}{\underset{i=1}{T}}\{\widetilde{\text{neg}}(a_i; \text{neg}(\nu))\}; \text{neg}(\nu)\right) \\ &= \widetilde{\text{neg}}\left(\overset{n}{\underset{i=1}{S}}\{\widetilde{\text{neg}}(a_i; \nu)\}; \nu\right),\end{aligned} \tag{4.12}$$

where $\widetilde{\text{neg}}(a; \nu)$ is a compromise operator expressed as follows:

$$\widetilde{\text{neg}}(a; \nu) = \text{neg}(\nu) \cdot \text{neg}(a) + \nu \cdot a. \tag{4.13}$$

Parameter $\nu \in [0, 1]$ occurring in formulas (4.12) and (4.13) decides whether operator $\tilde{H}(\mathbf{a}; \nu)$ is a t-norm (for $\nu = 0$) or a t-conorm (for $\nu = 1$).

The compromise operator (4.13) can be built e.g. on the basis of Zadeh negation (2.14). In this case it has the following form:

$$\widetilde{\mathrm{neg}_Z}(a; \nu) = (1 - \nu) \cdot (1 - a) + \nu \cdot a. \tag{4.14}$$

An example of the h-function constructed on the basis of Zadeh negation (2.14) and min/max-type triangular norms is an operator expressed as follows:

$$\begin{aligned}\tilde{H}_m(a_1, a_2; \nu) &= \widetilde{\mathrm{neg}_Z}\left(\min\left\{\widetilde{\mathrm{neg}_Z}(a_1; 1-\nu), \widetilde{\mathrm{neg}_Z}(a_2; 1-\nu)\right\}; 1-\nu\right) \\ &= \widetilde{\mathrm{neg}_Z}\left(\max\left\{\widetilde{\mathrm{neg}_Z}(a_1; \nu), \widetilde{\mathrm{neg}_Z}(a_2; \nu)\right\}; \nu\right).\end{aligned} \tag{4.15}$$

Another example of the h-function constructed on the basis of Zadeh negation (2.14) and algebraic triangular norms is an operator expressed as follows:

$$\begin{aligned}&\tilde{H}_a(a_1, a_2; \nu) = \widetilde{\mathrm{neg}_Z}\left(\widetilde{\mathrm{neg}_Z}(a_1; 1-\nu) \cdot \widetilde{\mathrm{neg}_Z}(a_2; 1-\nu); 1-\nu\right) \\ &= \widetilde{\mathrm{neg}_Z}\left(1 - \left(1 - \widetilde{\mathrm{neg}_Z}(a_1; 1-\nu)\right) \cdot \left(1 - \widetilde{\mathrm{neg}_Z}(a_2; 1-\nu)\right); \nu\right).\end{aligned} \tag{4.16}$$

On the basis of the h-function we can define switchable inference operator, which has the following form:

$$\tilde{I}(a_1, a_2; \nu) = \tilde{H}\left(\widetilde{\mathrm{neg}}(a_1; N(\nu)), a_2; \nu\right). \tag{4.17}$$

Parameter $\nu \in [0, 1]$ occurring in formula (4.17) decides whether for $\nu = 0$ operator $\tilde{I}(a_1, a_2; \nu)$ is a t-norm (a typical inference operator for Mamdani-type systems) or for $\nu = 1$ is an s-implication (a typical inference operator for logical-type systems).

An example of a switchable inference operator constructed on the basis of Zadeh negation (2.14) and min/max-type h-function of the form (4.15) is a switchable inference operator expressed as follows:

$$\begin{aligned}\tilde{I}_m(a_1, a_2; \nu) &= \widetilde{\mathrm{neg}_Z}\left(\min\left\{\begin{array}{c}\widetilde{\mathrm{neg}_Z}\left(\widetilde{\mathrm{neg}_Z}(a_1; 1-\nu); 1-\nu\right), \\ \widetilde{\mathrm{neg}_Z}(a_2; 1-\nu)\end{array}\right\}; 1-\nu\right) \\ &= \widetilde{\mathrm{neg}_Z}\left(\max\left\{\begin{array}{c}\widetilde{\mathrm{neg}_Z}\left(\widetilde{\mathrm{neg}_Z}(a_1; 1-\nu); \nu\right), \\ \widetilde{\mathrm{neg}_Z}(a_2; \nu)\end{array}\right\}; \nu\right).\end{aligned} \tag{4.18}$$

Analogously, an example of a switchable inference operator constructed on the basis of Zadeh negation (2.14) and the algebraic-type h-function of the form (4.16) is a switchable inference operator expressed as follows:

$$\tilde{I}_a(a_1, a_2; \nu) = \widetilde{\text{negz}}\left(\begin{array}{c}\widetilde{\text{negz}}\left(\widetilde{\text{negz}}(a_1; 1-\nu); 1-\nu\right)\cdot \\ \cdot\widetilde{\text{negz}}(a_2; 1-\nu); 1-\nu\end{array}\right) \\ = \widetilde{\text{negz}}\left(\begin{array}{c}1-\left(1-\widetilde{\text{negz}}\left(\widetilde{\text{negz}}(a_1; 1-\nu); 1-\nu\right)\right)\cdot \\ \cdot\left(1-\widetilde{\text{negz}}(a_2; 1-\nu)\right); \nu\end{array}\right). \tag{4.19}$$

The use of h-functions and h-implications is based on two typical cases:

- The value of the parameter ν is a real number from the unit range. Therefore, it can be selected e.g. in the learning process using gradient or population algorithm. In this variant, it is difficult to find an interpretation for the action of the operators (4.12), (4.13) and (4.17) when the value of the compromise parameter ν is different from 0 or 1. Of course, an action of the operator (4.12) can be interpreted as e.g. "more of an intersection of fuzzy sets than their sum," "more of a t-norm than t-conorm", etc. Similarly, an action of the operator (4.17) can be interpreted as e.g. "more of an inference of Mamdani-type than logical-type", "more of a t-norm than s-implication", etc. Then, an action of the operator (4.13) can be interpreted as e.g. "more of a linear function than negation". However, it does not change the fact that in this case it is difficult to give an unambiguous interpretation. Not only does this concern the operators defined in this section, but also other alternative operators present in the literature.
- The value of the parameter ν is an integer number from the set $\nu \in \{0, 1\}$. Therefore, it can be selected e.g. in a learning process using a gradient algorithm and properly defined range functions [43] or a population algorithm. This case seems to be of greater importance in the context of interpretable fuzzy systems design, because it ensures greater clarity of action of the operators used.

4.1.4 Hybrid Aggregation and Inference Operators

Hybrid aggregation and inference operators are often used in interpretable fuzzy systems designing. This section shows sample operators of this type, which combine advantages of the operators presented in Sects. 4.1.1–4.1.3. These operators are denoted as $\overset{\leftrightarrow}{\tilde{H}}{}^{*}(\mathbf{a}; \mathbf{w}, p, \nu)$ and $\overset{\leftrightarrow}{\tilde{I}}{}^{*}(a_1, a_2; w_1, w_2, p, \nu)$ and they are used in designing interpretable fuzzy systems discussed in the following chapters of this book.

Switchable parametrized aggregation operator with weights of arguments can be written as follows:

$$\overset{\leftrightarrow}{\tilde{H}}{}^{*}(\mathbf{a}; \mathbf{w}, p, \nu) = \overset{\leftrightarrow}{\tilde{H}}\left(\begin{array}{c}\widetilde{\arg}(a_1; w_1, p, \nu), \ldots, \widetilde{\arg}(a_n; w_n, p, \nu); \\ p, \nu\end{array}\right), \tag{4.20}$$

where $\overset{\leftrightarrow}{\tilde{H}}(\mathbf{a}; p, \nu)$ is an operator resulting from the connection of parameterized triangular norms (see Sect. 4.1.1) with triangular norms with weights of arguments (see Sect. 4.1.2) and switchable triangular norms (see Sect. 4.1.3). The operator $\widetilde{\arg}(a_i; w_i, p, v)$ in the formula (4.20) is a switchable argument of the following form:

$$\widetilde{\arg}(a; w, p, v) = \overset{\leftrightarrow}{\tilde{H}}\begin{pmatrix} a, \widetilde{\text{neg}}(w; v); \\ p, \text{neg}(v) \end{pmatrix}. \tag{4.21}$$

In formulas (4.20) and (4.21) there are h-functions, which do not have to be based on the same type of triangular norms, especially not on dual triangular norms.

On the basis of a switchable and parametrized aggregation operator with weights of arguments of the form (4.20) and its switchable argument of the form (4.21), switchable and parametrized inference operator with weights of arguments can be defined. It has the following form:

$$\overset{\leftrightarrow *}{\tilde{I}}(a_1, a_2; w_1, w_2, p, v) = \overset{\leftrightarrow}{\tilde{H}}\begin{pmatrix} \widetilde{\text{neg}}(\widetilde{\arg}(a_1; w_1, p, v); \text{neg}(v)), \\ \widetilde{\arg}(a_2; w_2, p, v); \\ p, v \end{pmatrix}. \tag{4.22}$$

In formulas (4.21) and (4.22) there are h-functions which also do not have to be based on the same type of triangular norms, especially on dual triangular norms (similar to the case of formulas (4.20) and (4.21)).

4.2 Precise Defuzzification Operators

As mentioned in Sect. 2.1, defuzzification aims at a transition from fuzzy sets to real (sharp) output signals [18, 33, 53]. The way in which defuzzification is carried out has a large influence on system accuracy. In practice in defuzzification operators used, the number of discretization points often equals the number of rules. This situation occurs in the systems discussed in Sect. 2.1. It is not beneficial because:

- Output fuzzy sets do not have to have one maximum. However, in the method described by formula (2.8), it is assumed that discretization takes place at the points in which output fuzzy sets reach their individual maximum.
- The number of discretization points of an output fuzzy set does not have to be equal to the number of rules. This is particularly important in a situation where a small number of fuzzy rules describes a problem well, but the corresponding small number of discretization points is insufficient to achieve the expected accuracy. Then, increasing the precision of a defuzzification operator (increasing the number of discretization points in particular) results in unnecessary increasing of the number of fuzzy rules.
- This prevents reduction of rules. Linking the number of rules with the number of discretization points causes that any removal of a rule deteriorate the defuzzifica-

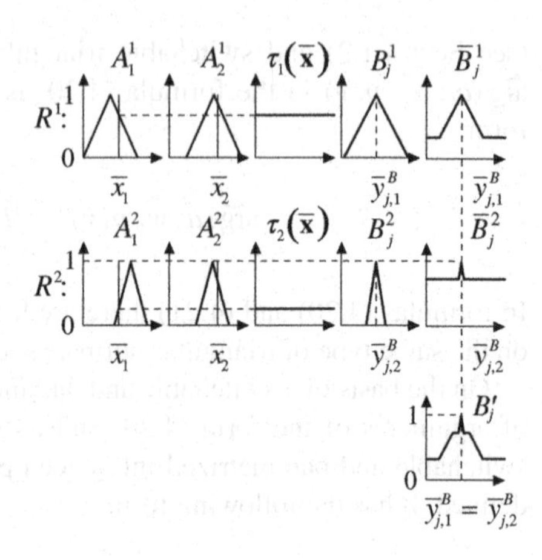

Fig. 4.1 Illustration of a sample defuzzification, in which the centers of output fuzzy sets B_j^1 i B_j^2 and the centers of fuzzy sets $\bar{B}_j^1$ i $\bar{B}_j^2$ from the rules overlap

tion accuracy. Thus, despite the fact that a rule is not needed in the system (e.g. it is not activated), it cannot be reduced, because it significantly deteriorates accuracy.

Figure 4.1 presents a sample schema of a logical-type inference, in which the centers of output fuzzy sets B_j^1 and B_j^2 overlap. Therefore, the centers of fuzzy inferences described by the sets $\bar{B}_j^1$ and $\bar{B}_j^2$, also overlap. In the case of defuzzification using method (2.8), it causes discretization of the resulting fuzzy set B_j' only in a single point. Then, regardless of the number of rules N used in the system, the response of the system is independent of the form of fuzzy set B_j' ($\bar{y}_j = \bar{y}_{j,1}^B = \cdots = \bar{y}_{j,N}^B$). Such system cannot work properly.

The idea for precise defuzzification operators could involve making the way in which defuzzification is carried out independent from the form of the rule base. This involves determination of the number of discretization points (their initialization) and their spacing. Making the number of discretization points independent from the number of rules leads to new discretization formulas. The new version of COA defuzzification method presented in Sect. 2.1 and described by formula (2.8) takes the following form:

$$\bar{y}_j = \frac{\sum\limits_{r=1}^{R_j} \bar{y}_{j,r}^{\text{def}} \cdot \mu_{B_j'}\left(\bar{y}_{j,r}^{\text{def}}\right)}{\sum\limits_{r=1}^{R_j} \mu_{B_j'}\left(\bar{y}_{j,r}^{\text{def}}\right)}, \tag{4.23}$$

where $\bar{y}_{j,r}^{\text{def}}$ ($j = 1, \ldots, m$, $r = 1, \ldots, R_j$) are discretization points and R_j is the number of discretization points for the output j [15]. In most applications, for the sake of simplicity it can be assumed that the number of discretization points is the same for each output of the system, i.e. $R_1 = \cdots = R_m = R$. However, it may happen that for certain system outputs it is a good idea to increase the number of discretization points. It results from the fact that for $R > N$ the formula (4.23) maps not-discretized defuzzification operator of the form (2.7) better. It is also important that increasing the number of discretization points does not reduce fuzzy system interpretability. This is very important from the point of view of interpretable fuzzy systems designing.

4.3 Flexible-Type Fuzzy Systems

Flexible-type fuzzy systems are the ones which use solutions aimed at increasing precision of aggregation, inference and defuzzification operators in order to obtain expected accuracy with the smallest rule base possible.

In the following part of this book systems constructed on the basis of Mamdani-type systems (described in Sect. 2.1.1) and logical-type systems (described in Sect. 2.1.2) are presented. In the notation of these systems standard triangular norms have been replaced by switchable, parametrized aggregation operators with weights of arguments of the form (4.20). Moreover, inference operators have been replaced by switchable, parametrized inference operators with weights of arguments of the form (4.22). This allows us to extend the notation of fuzzy rules of the form (2.2) to the following form:

$$R^k : \left[\left(\begin{array}{c} \text{IF } \left(x_1 \text{ is } A_1^k\right) \left|w_{1,k}^A \text{ AND } \ldots \text{ AND } \left(x_n \text{ is } A_n^k\right) \right|w_{n,k}^A \\ \text{THEN } \left(y_1 \text{ is } B_1^k\right) |w_{1,k}^B, \ \ldots, \left(y_m \text{ is } B_m^k\right) |w_{m,k}^B \end{array} \right) \left|w_k^{\text{rule}}\right. \right], \tag{4.24}$$

where $w_{i,k}^A \in [0,1]$ $(i = 1, \ldots, n,\ k = 1, \ldots, N)$ are weights of input fuzzy sets, $w_{j,k}^B \in [0,1]$ $(j = 1, \ldots, m,\ k = 1, \ldots, N)$ are weights of output fuzzy sets and $w_k^{\text{rule}} \in [0,1]$ $(k = 1, \ldots, N)$ are weights of fuzzy rules.

In this flexible-type fuzzy system the way of defuzzification has also changed. It uses the formula (4.23) instead of the formula (2.8). Membership functions of fuzzy sets B_j' appearing in this formula are expressed by operators of the form (4.20) and they take the following form:

$$\begin{aligned} \mu_{B_j'}\left(\bar{y}_{j,r}^{\text{def}}\right) &= \overset{\leftrightarrow}{\tilde{H}}^{*} \left(\begin{array}{c} \mu_{\bar{B}_j^1}\left(\bar{y}_{j,r}^{\text{def}}\right), \ldots, \mu_{\bar{B}_j^N}\left(\bar{y}_{j,r}^{\text{def}}\right); \\ w_1^{\text{rule}}, \ldots, w_N^{\text{rule}}, p_k^{\text{agr}}, 1-\nu \end{array} \right) \\ &= \overset{N}{\underset{k=1}{\overset{\leftrightarrow}{\tilde{H}}^{*}}} \left(\begin{array}{c} \mu_{\bar{B}_j^k}\left(\bar{y}_{j,r}^{\text{def}}\right); \\ w_k^{\text{rule}}, p_k^{\text{agr}}, 1-\nu \end{array} \right), \end{aligned} \tag{4.25}$$

where p_k^{agr} is a shape parameter of the used aggregation operator for the rule k and ν is an inference-type parameter, whose interpretation is described in Sects. 4.1.3 and 4.1.4.

Membership functions of fuzzy sets $\bar{B}_j^k$ in formula (4.25) are expressed by the operators of the form (4.22) and they take the following form:

$$\mu_{\bar{B}_j^k}\left(\bar{y}_{j,r}^{\mathrm{def}}\right) = \overset{\leftrightarrow}{\tilde{I}}^{*}\left(\begin{array}{c}\tau_k(\bar{\mathbf{x}}), \mu_{B_j^k}\left(\bar{y}_{j,r}^{\mathrm{def}}\right); \\ 1, w_{j,k}^{B}, p_k^{\mathrm{inf}}, \nu\end{array}\right), \tag{4.26}$$

where p_k^{inf} is the shape parameter of the used inference operator for the rule k.

Activity levels of rules $\tau_k(\bar{\mathbf{x}})$ in formula (4.26) are expressed by the operators of the form (4.20) and they take the following form:

$$\begin{aligned}\tau_k(\bar{\mathbf{x}}) &= \overset{\leftrightarrow}{\tilde{H}}^{*}\left(\begin{array}{c}\mu_{A_1^k}(\bar{x}_1), \ldots, \mu_{A_n^k}(\bar{x}_n); \\ w_{1,k}^{A}, \ldots, w_{n,k}^{A}, p_k^{\tau}, 0\end{array}\right) \\ &= \overset{n}{\underset{i=1}{\overset{\leftrightarrow}{\tilde{H}}^{*}}}\left(\begin{array}{c}\mu_{A_i^k}(\bar{x}_i); \\ w_{i,k}^{A}, p_k^{\tau}, 0\end{array}\right),\end{aligned} \tag{4.27}$$

where p_k^{τ} is the shape parameter of the used aggregation operator for the rule k. The operator of the form (4.27) is a t-norm because its last parameter is equals 0.

Taking into account formulas (4.25), (4.26) and (4.27) in the assumed defuzzification formula (4.23), we obtain dependency describing the value of the output signal $\bar{y}_j$ for the flexible-type system:

$$\bar{y}_j = \frac{\sum\limits_{r=1}^{R_j} \bar{y}_{j,r}^{\mathrm{def}} \cdot \overset{N}{\underset{k=1}{\overset{\leftrightarrow}{\tilde{H}}^{*}}}\left(\begin{array}{c}\overset{\leftrightarrow}{\tilde{I}}^{*}\left(\begin{array}{c}\overset{n}{\underset{i=1}{\overset{\leftrightarrow}{\tilde{H}}^{*}}}\left(\begin{array}{c}\mu_{A_i^k}(\bar{x}_i); \\ w_{i,k}^{A}, p_k^{\tau}, 0\end{array}\right), \mu_{B_j^k}\left(\bar{y}_{j,r}^{\mathrm{def}}\right); \\ 1, w_{j,k}^{B}, p_k^{\mathrm{inf}}, \nu\end{array}\right); \\ w_k^{\mathrm{rule}}, p_k^{\mathrm{agr}}, 1-\nu\end{array}\right)}{\sum\limits_{r=1}^{R_j} \overset{N}{\underset{k=1}{\overset{\leftrightarrow}{\tilde{H}}^{*}}}\left(\begin{array}{c}\overset{\leftrightarrow}{\tilde{I}}^{*}\left(\begin{array}{c}\overset{n}{\underset{i=1}{\overset{\leftrightarrow}{\tilde{H}}^{*}}}\left(\begin{array}{c}\mu_{A_i^k}(\bar{x}_i); \\ w_{i,k}^{A}, p_k^{\tau}, 0\end{array}\right), \mu_{B_j^k}\left(\bar{y}_{j,r}^{\mathrm{def}}\right); \\ 1, w_{j,k}^{B}, p_k^{\mathrm{inf}}, \nu\end{array}\right); \\ w_k^{\mathrm{rule}}, p_k^{\mathrm{agr}}, 1-\nu\end{array}\right)}. \tag{4.28}$$

In the following part of the book mainly systems based on the Gaussian membership function are used. The values of this function tend asymptotically to 0 but do not reach it, which is not advantageous in the context of our considerations on interpretability. However, this function is of a great practical importance in modelling of real processes, because it reflects well their true character (i.e. their course), which has been a decisive feature for using it. However, our considerations are general in character and could relate to any other membership function. The Gaussian function is described as follows:

$$\mu_G(x; \bar{x}, \sigma) = \exp\left(-\left(\frac{x-\bar{x}}{\sigma}\right)^2\right), \tag{4.29}$$

where $\bar{x}$ is the center of the function and σ is its width.

It is easy to notice that the system of the form (4.28) implements Mamdani-type inference (considered in Sect. 2.1.1) for $\nu = 0$ and logical-type inference (considered in Sect. 2.1.2) for $\nu = 1$. Moreover, it contains the following parameters:

- Input fuzzy sets parameters A_i^k $(i = 1, \ldots, n, k = 1, \ldots, N)$. They are included in the membership function $\mu_{A_i^k}(\cdot)$. Number and interpretation of these parameters depend on the assumed type of sets. If we use Gaussian membership function of the form (4.29), then each fuzzy set A_i^k is described by two parameters: center $\bar{x}_{i,k}^A$ and width $\sigma_{i,k}^A$.
- Output fuzzy sets parameters B_j^k $(j = 1, \ldots, m, k = 1, \ldots, N)$. They are included in the membership function $\mu_{B_j^k}(\cdot)$. The number and interpretation of these parameters depend on the assumed type of sets. If we use Gaussian membership function of the form (4.29), then each fuzzy set B_j^k is described by two parameters: the center $\bar{x}_{j,k}^B$ and the width $\sigma_{j,k}^B$.
- Weights of importance $w_{i,k}^A \in [0, 1]$ $(i = 1, \ldots, n, k = 1, \ldots, N)$ of input fuzzy sets A_i^k.
- Weights of importance $w_{j,k}^B \in [0, 1]$ $(j = 1, \ldots, m, k = 1, \ldots, N)$ of output fuzzy sets B_j^k.
- Weights of importance $w_k^{\text{rule}} \in [0, 1]$ $(k = 1, \ldots, N)$ of fuzzy rules.
- Shape parameters p_k^τ $(k = 1, \ldots, N)$ of parameterized aggregation operators of membership function of input fuzzy sets A_i^k. The range of values of parameters p_k^τ depends on the type of parameterized triangular norms used for the construction of aggregation operators.
- Shape parameters p_k^{inf} $(k = 1, \ldots, N)$ of parameterized inference operators. The range of values of parameters p_k^{inf} depends on the type of parameterized triangular norms used for the construction of aggregation operators.
- Shape parameters p_k^{agr} $(k = 1, \ldots, N)$ of parameterized aggregation operators of the membership function of output fuzzy sets $\bar{B}_j^k$. The range of values of parameters p_k^{agr} depends on the type of parameterized triangular norms used for the construction of aggregation operators.
- Discretization points $\bar{y}_{j,r}^{\text{def}}$ $(j = 1, \ldots, m, r = 1, \ldots, R_j)$ of fuzzy set B_j'.

Values of the parameters of the fuzzy system of the form (4.28) are usually selected using methods described in Sect. 2.2 in order to achieve satisfactory system accuracy. In interpretable fuzzy systems designing an additional issue is such impact on the values of these parameters so as to make their interpretation possible. In this case using interpretability criteria may prove helpful. In Sect. 3.4 the basic interpretability criteria of fuzzy systems are described. They relate to the first two types of the parameters, i.e. input fuzzy sets parameters A_i^k and output fuzzy sets parameters B_j^k. Then, the next section contains a description of interpretability criteria related to the other types of system parameters (4.28).

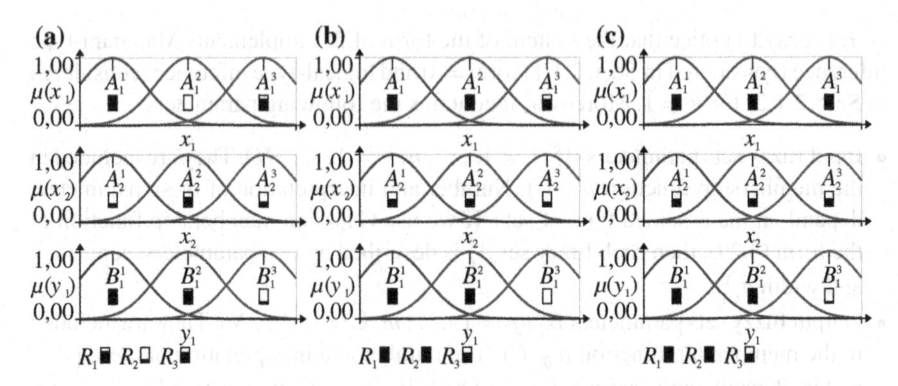

Fig. 4.2 Three sample cases illustrating the problem of the readability of weights from the rule base of the form (4.24): **a** unfavorable, **b** intermediate, **c** favorable (good weight readability, a low criterion value). *Black* rectangles mean weight value equal to 1.0 while *white* rectangles mean weight value equal to 0.0

4.4 Extended Interpretability Criteria of Fuzzy Systems

In this section (and in Sect. 3.4) we describe interpretability criteria which can be used when designing interpretable fuzzy systems. The criteria described in this section relate to the flexibility aspects of the fuzzy system of the form (4.28) discussed in Sects. 4.1–4.3. So far this subject has mainly been raised in our previous work [16]. Several individual solutions have also appeared in the works of other authors (e.g. [34]).

An additional advantage of the criteria considered in this section is a more comprehensive approach to fuzzy systems interpretability, without focusing only on the "fuzzy set" and "fuzzy rule" notions. Extended interpretability criteria of fuzzy systems can include:

- Complexity criterion (Fig. 3.3). This criterion has been previously classified as a basic one, but it may also take into account the system components (e.g. discretization points) which relate to the impact on fuzzy systems interpretability by appropriate selection of their structure.
- Criterion for assessing the readability of weight values in the rule base of the form (4.24) of the system (4.28) (Fig. 4.2). The preferred weight values are close to the values of the set {0.0, 0.5, 1.0}. Then, we can easily label them as "not important", "important" and "very important". These labels relate to the fuzzy system elements associated with weights (i.e. its fuzzy sets or rules). Using a function which promotes the values 0.0, 0.5 and 1.0 could be very helpful in formulating the criterion. This function can be defined as follows:

Table 4.3 A list of selected values of the parameter of Dombi-type parameterized triangular norms of the form (4.1), for which parameter their shape approximates the shape of typical and non-parametrized triangular norms

Parameter value	Non-parameterized norm	Similarity level
$p \to \infty$	Minimum/maximum	Full
$p = 0.71$	Łukasiewicz	High
$p = 0.43$	Algebraic	High
$p = 0.00$	Drastic	Full

$$\mu_w(x) = \begin{cases} \frac{a-x}{a} & \text{for } x \in [0, a] \\ \frac{x-a}{b-a} & \text{for } x \in (a, b] \\ \frac{c-x}{c-b} & \text{for } x \in (b, c] \\ \frac{x-c}{1-c} & \text{for } x \in (c, 1], \end{cases} \tag{4.30}$$

where $a = 0.25$, $b = 0.50$, and $c = 0.75$.

- Criterion for assessing the readability of shape parameters' values of parametrized triangular norms. The preferred weight values are the ones, for which parametrized norms provide the best possible match for their non-flexible counterparts. It is assumed that the operation of non-flexible norms is more readable than their parameterized counterparts. Therefore, this approach concerns interpretability at the level of operators. The expected values for Dombi norms of the form (4.1) are exemplified in Table 4.3. Using the function promoting expected values could be helpful in formulating the criterion. The function for the values presented in Table 4.3 can be defined as follows (Fig. 4.3):

$$\mu_p(x) = \begin{cases} \frac{a-x}{a} & \text{for } x \in [0, a] \\ \frac{x-a}{b-a} & \text{for } x \in (a, b] \\ \frac{c-x}{c-b} & \text{for } x \in (b, c] \\ \frac{x-c}{d-c} & \text{for } x \in (c, d] \\ \frac{e-x}{e-d} & \text{for } x \in (d, e] \\ \frac{x-e}{f-e} & \text{for } x \in (e, f], \end{cases} \tag{4.31}$$

where $a = 0.21$, $b = 0.43$, $c = 0.57$, $d = 0.71$, $e = 0.85$, and $f = 10.00$.

- Criterion for assessing the readability of inference model. A readable inferencing scheme is the one which the best corresponds to the Mamdani- or logical-type inference (in the context of a flexible system described by the formula (4.28)). Using a function formulated in a similar way to the one in the case of the weight readability assessing criterion could be helpful in formulating the criterion.
- Criterion for assessing the readability of discretization points (Fig. 4.4). Favorably distributed discretization points are close to the centers of output fuzzy sets and within them.

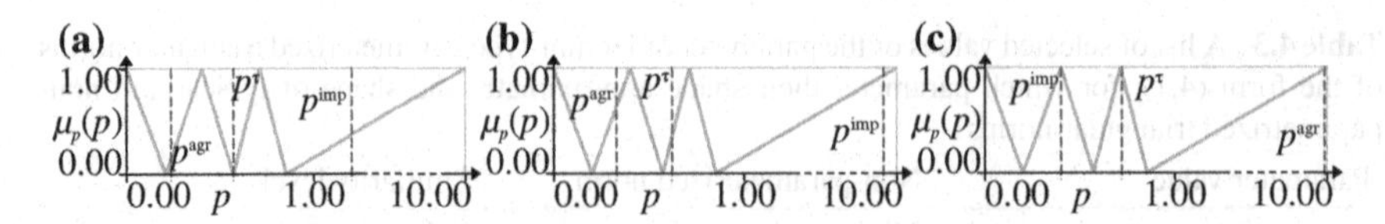

Fig. 4.3 Three sample cases illustrating the problem of the readability of shape parameters of Dombi-type norms (in the sense of similarity to the values presented in Table 4.3): **a** unfavorable, **b** intermediate, **c** favorable (good shape parameters readability, a low criterion value)

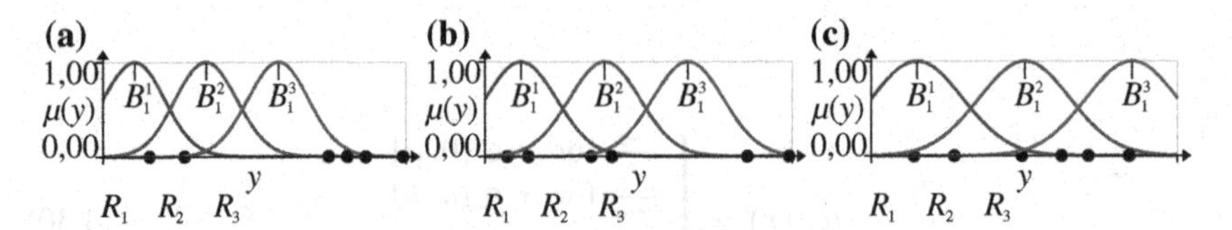

Fig. 4.4 Three sample cases illustrating the problem of discretization points readability: **a** unfavorable, **b** intermediate, **c** favorable (properly distributed discretization points, a low criterion value). Discretization points are denoted as *black circles*

Sample formulas defining the interpretability criteria mentioned above are presented in Chap. 6.

In interpretable fuzzy systems designing we can distinguish two different approaches to enforcing interpretability, which have been already mentioned in Sect. 4.1.3 in the context of using an h-function and an h-implication. They are the following approaches:

- An approach based on the ranges of values. It assumes that the real parameters of the system (4.28) can take any real values from their domain. In this case enforcing interpretability might involve the use of appropriate interpretability criteria in order to select a system that meets specific requirements. However, it allows for a fairly limited scope for action. This approach is commonly used in the context of fuzzy systems learned by a gradient algorithm.
- An approach based on the sets of values. It assumes that the real parameters of the system (4.28) can take values from the dedicated set which has been selected earlier. In this case enforcing interpretability involves appropriate determination of values included in the sets of allowed values. It is simple in the case of parameters such as e.g. an inference-type parameter, weight of importance or shape parameters of parameterized triangular norms. It is also possible, e.g., in the case of discretization points and the input and output fuzzy sets parameters, which may be indicated by the index from a certain pool. This approach is easy to implement in the context of fuzzy systems learned by a population algorithm. It gives great opportunities for automatic selection of the fuzzy system structure.

4.5 Simulation Results

Comments on the simulations can be summarized as follows:

- The aim of the simulations was to show the impact of various aspects of fuzzy systems structure selection on their accuracy. It was assumed that a positive impact on accuracy also showed a beneficial impact on interpretability, because in such case expected accuracy could be achieved using a less number of fuzzy rules.
- A list of the simulation problems presented is shown in Table 4.4. They are sample problems from the field of approximation and classification.
- In the learning phase an evolutionary strategy (μ, λ) was used [14, 40] with $\mu = 500$ and $\lambda = 100$. The learning process took 1000 steps (generations).
- For the problems from the field of approximation each simulation was repeated 100 times. Evaluation of their effectiveness was performed using the RMSE error expressed by formula (2.18).
- For the problems from the field of classification the stratified cross-validation described in Sect. 2.3.2 was carried out. An evaluation of their effectiveness was performed using the percentage classification error expressed by formula (2.20).
- The simulation results presented in Tables 4.5, 4.6, 4.7, 4.8, 4.9, 4.10 and 4.11 relate to different variants of the fuzzy system: System1 is a non-flexible system of the form (2.8), System2 is a flexible system of the form (4.28) using a precise defuzzification operator (4.23), System3 is a flexible system of the form (4.28) using precise Dombi-type aggregation and inference operators of the form (4.1), System4 is a flexible system of the form (4.28) using precise aggregation and inference operators with weights of arguments of the forms (4.8) and (4.11), System5 is a flexible system of the type (4.28) with a precise hybrid-type aggregation and inference operators of the forms (4.20) and (4.22). In these systems for the tasks from the field of approximation Mamdani-type inference was used while for the tasks related to the classification the logical-type inference was used [43].
- The results concerning the accuracy of the non-flexible system of the form (2.8) and different variants of the flexible system of the form (4.28) are summarized in Tables 4.5, 4.6, 4.7 and 4.8.
- The results concerning the average, percentage increase of the accuracy of different flexible fuzzy systems of the form (4.28) in relation to the ones obtained for the base non-flexible system of the form (2.8) are summarized in Tables 4.9, 4.10 and 4.11.

The conclusions from the simulations can be summarized as follows:

- The right choice of the fuzzy system structure resulted in obtaining higher accuracy than in the case of the non-flexible system of the form (2.8). It was evident for the issues of approximation and classification alike (Tables 4.9, 4.10 and 4.11).
- The highest increase of accuracy in the System2-System5 group was achieved using a flexible system of the form (4.28) with a precise hybrid-type aggregation and inference operators of the form (4.20) and (4.22) (System5). The increase

Table 4.4 A list of the test problems used in the simulations

Item no.	Problem name	Number of input attributes	Number of output attributes	Number of sets	Problem type	Problem label
1	Concrete slump [54]	7	3	103	Approximation	CS
2	Nelson function [35]	2	1	128	Approximation	NF
3	Pima Indians diabetes [31]	8	2	768	Classification	PID
4	Wine recognition [31]	13	3	178	Classification	WC

Table 4.5 A list of the average RMSE errors for a non-flexible system of the form (2.8) and different variants of flexible system of the form (4.28) for the CS problem

System	$N = 2$	$N = 3$	$N = 4$	$N = 5$	$N = 6$	Average result
1	17.4040	15.6160	14.6347	15.0459	14.8570	15.5115
2	16.7144	14.9832	14.8350	14.3415	14.6178	15.0984
3	16.5510	14.7538	15.3262	13.9141	14.3390	14.9768
4	16.0596	15.6260	14.3761	14.9836	14.5227	15.1136
5	15.8481	14.6350	14.2013	14.0613	14.0007	14.5493

Table 4.6 A list of the average RMSE errors for a non-flexible system of the form (2.8) and different variants of a flexible system of the form (4.28) for the NF problem

System	$N = 2$	$N = 3$	$N = 4$	$N = 5$	$N = 6$	Average result
1	1.5142	1.3791	1.3915	1.3379	1.2759	1.3797
2	1.5117	1.3688	1.3386	1.3106	1.2719	1.3603
3	1.4837	1.3703	1.3392	1.3122	1.2715	1.3554
4	1.4733	1.3509	1.2658	1.2896	1.2652	1.3289
5	1.4060	1.3126	1.2738	1.2338	1.1949	1.2842

was only slightly more significant for the considered classification problems (Table 4.10) than those of the approximation (Table 4.9).

- The highest impact on the accuracy of System2-System4 systems had, respectively, using triangular norms with weights of arguments of the forms (4.8) and (4.11) (System4), using a precise defuzzification operator of the form (4.23) (System2), and using precise Dombi-type aggregation and inference operators of the forms (4.1) (System3).

Table 4.7 A list of the average RMSE errors for a non-flexible system of the form (2.8) and different variants of a flexible system of the form (4.28) for the PID problem

System	$N = 2$	$N = 3$	$N = 4$	$N = 5$	$N = 6$	Average result
1	25.57	25.59	24.56	24.12	23.59	24.69
2	25.39	25.66	24.68	23.68	23.87	24.66
3	25.53	25.13	24.87	24.12	23.81	24.69
4	23.95	23.37	23.11	22.90	23.03	23.27
5	22.44	23.81	22.79	21.80	22.20	22.61

Table 4.8 A list of the average RMSE errors for a non-flexible system of the form (2.8) and different variants of a flexible system of the form (4.28) for the WC problem

System	$N = 2$	$N = 3$	$N = 4$	$N = 5$	$N = 6$	Average result
1	11.76	10.00	9.41	10.59	10.88	10.53
2	9.41	8.91	7.06	7.35	8.82	8.31
3	8.82	7.65	8.82	11.18	11.76	9.65
4	7.65	7.46	7.12	6.62	7.09	7.19
5	6.24	7.13	6.27	6.66	8.01	6.86

Table 4.9 A list of the average percentage increase in the accuracy of a flexible system of the form (4.28) in relation to the accuracy of a non-flexible system of the form (2.8) for the approximation problems

System	$N = 2$	$N = 3$	$N = 4$	$N = 5$	$N = 6$	Average result
2	2.06	2.40	1.22	3.36	0.96	2.00
3	3.46	3.08	−0.49	4.72	1.91	2.54
4	5.21	0.99	5.40	2.01	1.54	3.03
5	8.04	5.55	5.71	7.16	6.06	6.50

Table 4.10 A list of the average percentage increase in the accuracy of a flexible system of the form (4.28) in relation to the accuracy of a non-flexible system of the form (2.8) for the classification problems

System	$N = 2$	$N = 3$	$N = 4$	$N = 5$	$N = 6$	Average result
2	10.34	5.30	12.26	16.18	8.84	10.59
3	12.58	12.66	2.50	–2.77	–4.54	4.08
4	20.67	17.01	15.12	21.26	18.61	18.54
5	29.59	17.81	20.32	23.37	16.12	21.44

Table 4.11 A list of the average percentage increase in the accuracy in relation to the accuracy obtained for the base system of the form (2.8) for the approximation and classification problems

System	$N=2$	$N=3$	$N=4$	$N=5$	$N=6$	Average result
2	6.20	3.85	6.74	9.77	4.90	6.29
3	8.02	7.87	1.01	0.98	–1.31	3.31
4	12.94	9.00	10.26	11.64	10.08	10.78
5	18.81	11.68	13.01	15.27	11.09	13.97

4.6 Summary

In this chapter different variants of the so-called precise aggregation and inference operators are presented and discussed. They were constructed on the basis of triangular norms with weights of arguments, parameterized triangular norms and switchable triangular norms. Hybrid aggregation and inference operators were proposed on the basis of these types of norms. In addition to the inference and aggregation operators, the so-called precise defuzzification operators were considered, which, among others, can make the inference mechanism to become independent of the defuzzification one. This creates many beneficial opportunities for various processes including interpretable fuzzy systems designing. They are also considered in Chaps. 5 and 6. In the following part of the chapter the flexible type systems are described. Their construction is based on using so-called precise operators. The systems were developed on the basis of non-flexible systems (Chap. 2). For non-flexible systems the extended interpretability criteria were proposed. They supplement the criteria described in Sect. 3.4. The final part of the chapter contains sample results obtained from the simulations carried out on the flexible-type systems. The most important conclusion from the simulations is that using precise operators has a positive impact on fuzzy system accuracy. It is a good basis for action aimed at minimizing the rule base. It also leads to the conclusion that expanding the rule base is not the only mechanism for increasing fuzzy system accuracy.

Some aspects of affecting interpretability of fuzzy systems by appropriate selection of their structure were partially outlined in our previous papers (see e.g. [41–43]).

References

1. Alcalá, R., Casillas, J., Cordón, O., González, A., Herrera, F.: A genetic rule weighting and selection process for fuzzy control of heating, ventilating and air conditioning systems. Eng. Appl. Artif. Intell. **18**, 279–296 (2005)
2. Alcala-Fdez, J., Alcala, R., Herrera, F.: A fuzzy association rule-based classification model for high-dimensional problems with genetic rule selection and lateral tuning. IEEE Trans. Fuzzy Syst. **19**, 857–872 (2011)

3. Alshomrani, S., Bawakid, A., Shim, S.O., Fernández, A., Herrera, F.: A proposal for evolutionary fuzzy systems using feature weighting: dealing with overlapping in imbalanced datasets. Knowl. Based Syst. **73**, 1–17 (2015)
4. Alsina, C., Maurice, F., Schweizer, B.: Associative Functions: Triangular Norms And Copulas. WSPC (2006)
5. Anooj, P.K.: Clinical decision support system: risk level prediction of heart disease using weighted fuzzy rules. J. King Saud Univ. Comput. Inf. Sci. **24**, 27–40 (2012)
6. Box, G.E.P., Jenkins, G.M., Reinsel, G.C.: Time Series Analysis: Forecasting and Control. Wiley, New Jersey (2008)
7. Chang, L., Zhou, Z.J., You, Y., Yang, L., Zhou, Z.: Belief rule based expert system for classification problems with new rule activation and weight calculation procedures. Inf. Sci. **336**, 75–91 (2016)
8. Chen, S.M., Chang, Y.C.: Weighted fuzzy rule interpolation based on GA-based weight-learning techniques. IEEE Trans. Fuzzy Syst. **19**, 729–744 (2011)
9. Chen, S.M., Hsin, W.C.: Weighted fuzzy interpolative reasoning based on the slopes of fuzzy sets and particle swarm optimization techniques. IEEE Trans. Cybern. **45**, 1250–1261 (2015)
10. Chen, S.M., Huang, C.M.: Generating weighted fuzzy rules from relational database systems for estimating values using genetic algorithms. IEEE Trans. Fuzzy Syst. **11**, 495–506 (2003)
11. Chen, S.M., Ko, Y.K., Chang, Y.C., Pan, J.S.: Weighted fuzzy interpolative reasoning based on weighted increment transformation and weighted ratio transformation techniques. IEEE Trans. Fuzzy Syst. **17**, 1412–1427 (2009)
12. Chen, S.M., Lee, L.W., Shen, V.R.L.: Weighted fuzzy interpolative reasoning systems based on interval type-2 fuzzy sets. Inf. Sci. **248**, 15–30 (2013)
13. Cheng, S.H., Chen, S.M., Chen, C.L.: Fuzzy interpolative reasoning based on ranking values of polygonal fuzzy sets and automatically generated weights of fuzzy rules. Inf. Sci. **325**, 521–540 (2015)
14. Cpałka, K.: On evolutionary designing and learning of flexible neuro-fuzzy structures for nonlinear classification. Nonlinear Anal. Ser. A Theory Methods Appl Elsevier **71**, 1659–1672 (2009)
15. Cpałka, K.: A new method for design and reduction of neuro-fuzzy classification systems. IEEE Trans. Neural Netw. **20**, 701–714 (2009)
16. Cpałka, K., Łapa, K.: Przybył, Zalasiński, M.: A new method for designing neuro-fuzzy systems for nonlinear modelling with interpretability aspects. Neurocomputing **135**, 203–217 (2014)
17. Cpałka, K., Rebrova, O., Nowicki, R., Rutkowski, L.: On design of flexible neuro-fuzzy systems for nonlinear modelling. Int. J. Gen. Syst. **42**, 706–720 (2013)
18. De Oliveira, J.: A set-theoretical defuzzification method. Fuzzy Sets Syst. **76**, 63–71 (1995)
19. Dymowa, L.: Soft Computing in Economics and Finance. Springer, Heidelberg (2011)
20. Farahbod, F., Eftekhari, M.: Comparsion of different T-norm operators in classification problems. Int. J. Fuzzy Logic Syst. **2**, 33–41 (2012)
21. Gaxiola, F., Melin, P., Valdez, F., Castro, J.R., Castillo, O.: Optimization of type-2 fuzzy weights in backpropagation learning for neural networks using GAs and PSO. Appl. Soft Comput. **38**, 860–871 (2016)
22. Ishibuchi, H.: Rule weight specification in fuzzy rule-based classification systems. IEEE Trans. Fuzzy Syst. **13**, 428–436 (2005)
23. Ishibuchi, H., Nakashima, T.: Effect of rule weights in fuzzy rule-based classification systems. IEEE Trans. Fuzzy Syst. **9**, 506–515 (2001)
24. Ishibuchi, H., Yamamoto, T.: Rule weight specification in fuzzy rule-based classification systems. IEEE Trans. Fuzzy Syst. **13**, 428–435 (2005)
25. Jahromi, M.Z., Taheri, M.: A proposed method for learning rule weights in fuzzy rule-based classification systems. Fuzzy Sets Syst. **159**, 449–459 (2008)
26. Kauers, M., Pillwein, V., Saminger-Platz, S.: Dominance in the family of Sugeno-Weber t-norms. Fuzzy Sets Syst. **181**, 74–87 (2011)
27. Klement, E.P., Mesiar, R. (eds.): Logical, Algebraic, Analytic and Probabilistic Aspects of Triangular Norms. Elsevier, Amsterdam (2005)

28. Klement, E.P., Mesiar, R., Pap, E.: Triangular Norms. Springer, Heidelberg (2000)
29. Klement, E.P., Mesiar, R., Pap, E.: Triangular norms. Position paper II: general constructions and parameterized families. Fuzzy Sets Syst. **145**, 411–438 (2004)
30. Ko, J.W., Lee, W.I., Park, P.: Stabilization for Takagi-Sugeno fuzzy systems based on partitioning the range of fuzzy weights. Automatica **48**, 970–973 (2012)
31. Lichman, M.: UCI Machine Learning Repository. Irvine, CA: University of California, School of Information and Computer Science (2013). http://archive.ics.uci.edu/ml
32. Lin, C.T., Lu, Y.C.: A neural fuzzy system with fuzzy supervised learning. IEEE Trans. Syst. Man Cybern. Part B (Cybern.) **26**, 744–763 (1996)
33. Mahdiani, H.R., Banaiyan, A., Javadi, M.H.S., Fakhraie, S.M., Lucas, C.: Defuzzification block: new algorithms, and efficient hardware and software implementation issues. Eng. Appl. Artif. Intell. **26**, 162–172 (2013)
34. Mencar, C., Castellano, G., Fanelli, A.M.: Some fundamental interpretability issues in fuzzy modeling. In: Proceedings of the Joint 4th Conference of the European Society for Fuzzy Logic and Technology (EUSFLAT-LFA), pp. 100–105 (2005)
35. Nelson, W.: Analysis of performance-degradation data. IEEE Trans. Reliab. **2**, 149–155 (1981)
36. Pedrycz, W.: Fuzzy Sets Engineering. CRC Press, Florida (1995)
37. Riid, A., Rüstern, E.: Interpretability, interpolation and rule weights in linguistic fuzzy modeling. In: Fanelli, A.M., Pedrycz, P.A., Petrosino, A. (eds.) Fuzzy Logic and Applications: Proceedings of the 9th International Workshop (WILF 2011), pp. 91–98. Springer, Heidelberg (2011)
38. Riid, A., Rüstern, E.: Adaptability, interpretability and rule weights in fuzzy rule-based systems. Inf. Sci. **257**, 301–312 (2014)
39. Rutkowski, L.: Flexible Neuro-Fuzzy Systems: Structures, Learning and Performance Evaluation. Springer, Heidelberg (2004)
40. Rutkowski, L.: Computational Intelligence. Springer, Heidelberg (2010)
41. Rutkowski, L., Cpałka, K.: A neuro-fuzzy controller with a compromise fuzzy reasoning. Control Cybern. **31**, 297–308 (2002)
42. Rutkowski, L., Cpałka, K.: Flexible neuro-fuzzy systems. IEEE Trans. Neural Netw. **14**, 554–574 (2003)
43. Rutkowski, L., Cpałka, K.: Designing and learning of adjustable quasi-triangular norms with applications to neuro-fuzzy systems. IEEE Trans. Fuzzy Syst. **13**, 140–151 (2005)
44. Schneider, M., Shnaider, E., Kandel, A.: Applications of the negation operator in fuzzy production rules. Fuzzy Sets Syst. **34**, 293–299 (1990)
45. Simiński, K.: Rule weights in a neuro-fuzzy system with a hierarchical domain partition. Int. J. Appl. Math. Comput. Sci. **20**, 337–347 (2010)
46. Sinova, B., Gil, M.A., López, M.T., Van Aelst, S.: A parameterized metric between fuzzy numbers and its parameter interpretation. Fuzzy Sets Syst. **245**, 101–115 (2014)
47. Sudkamp, T., Hammell, R.J.: Interpolation, completion, and learning fuzzy rules. IEEE Trans. Syst. Man Cybern. **24**, 332–342 (1994)
48. Troiano, L., Rodríguez-Muñiz, L.J., Marinaro, P., Díaz, I.: Statistical analysis of parametric t-norms. Inf. Sci. **257**, 138–162 (2014)
49. Türksen, B.: An Ontological and Epistemological Perspective of Fuzzy Set Theory. Elsevier Science, Amsterdam (2005)
50. Wang, X.Z., Dong, C.R., Fan, T.G.: Training T-S norm neural networks to refine weights for fuzzy if-then rules. Neurocomputing **70**, 2581–2587 (2007)
51. Wang, J., Shi, P., Peng, H., Pérez-Jiménez, M.J., Wang, T.: Weighted fuzzy spiking neural P systems. IEEE Trans. Fuzzy Syst. **21**, 209–220 (2013)
52. Weber, S.: A general concept of fuzzy connectives, negations and implications based on t-norms and t-conorms. Fuzzy Sets Syst. **11**, 115–134 (1983)
53. Yager, R.R., Filev, D.: On the issue of defuzzification and selection based on a fuzzy set. Fuzzy Sets Syst. **55**, 255–271 (1993)
54. Yeh, I.C.: Modeling slump flow of concrete using second-order regressions and artificial neural networks. Cem. Concr. Compos. **29**, 474–480 (2007)

Chapter 5
Interpretability of Fuzzy Systems Designed in the Process of Gradient Learning

Supervised learning of fuzzy systems can be used to select their structure and parameters. There are few approaches to supervised learning and one of them is gradient learning (Sect. 2.2). Impact on interpretability of fuzzy systems designed by gradient learning is a complex issue. It involves operations related to initialization, choice of a learning algorithm, realization of a learning process and also the final tuning of a system. Inappropriate actions in any of these phases will reduce system accuracy, might lengthen a learning process or make it impossible to reach an assumed threshold value as a result of a learning error. Then, decreasing system accuracy might necessitate an increase in the number of fuzzy rules.

The issue of fuzzy systems learning with the use of a gradient algorithm requires an adjustment of the learning algorithm to the system structure used. This is due to the close integration of formulas describing a given learning algorithm with the type of a fuzzy system. Gradient learning also has other limitations, which might include: limited opportunities in the field of automatic selection of the system structure (e.g. number of inputs, rules and fuzzy sets), limited opportunities in the field of automatic selection of the system structure components (e.g. type of aggregation and inference operators used), etc. These limitations make us seek supporting solutions, which might include algorithms of reduction, merging and automatic final tuning. They are all discussed in this chapter.

This chapter describes the issue of fuzzy systems initialization (Sect. 5.1), different aspects of fuzzy systems gradient learning (Sect. 5.2), the issue of fuzzy system final tuning (Sect. 5.3) and fuzzy system reduction algorithms (Sects. 5.4 and 5.5). The final part of the chapter contains exemplary simulation results (Sect. 5.6), a summary (Sect. 5.7) and bibliography.

K. Cpałka, *Design of Interpretable Fuzzy Systems*, Studies in Computational Intelligence 684, DOI 10.1007/978-3-319-52881-6_5

5.1 Initialization of Fuzzy Systems

Initialization of fuzzy systems designed by gradient learning is based on preliminary determination of their parameters values before the start of a learning process. Methods frequently used for initialization might include:

- Initialization of system parameters by using constant values.
- Initialization of system parameters by using random values. In this method the range of a draw should correspond to the range of initialized parameters.
- Even initialization of system parameters. In this method parameter values should cover evenly the entire domain of initialized parameters.
- Initialization of system parameters in such a way that their values cover significant areas within the field of initialized parameters. This approach often uses unsupervised learning, which might include different data clustering techniques (see e.g. [1, 15, 55]).

An interesting technique of fuzzy set parameters initialization is, as mentioned above, using possibilities offered by data clustering techniques [8, 56]. It is generally understood that data clustering is meant to divide the set of input data into a specified number of groups on the basis of an adopted similarity measure. Manually performed clustering is usually difficult because the number of clusters and input sets could be large. Many similarity measures can be used in a clustering process, e.g. the typical Euclidean measure. A selected similarity measure (e.g. the Manhattan measure) is often adapted to the specifics of a considered problem. Regardless of a chosen similarity measure, a division achieved using data clustering methods should be characterized by two features, i.e. heterogeneity of clusters and homogeneity of the data in clusters. There are three basic types of clustering methods:

- In hard methods it is assumed that each input object can only belong to one group.
- In fuzzy methods it is assumed that each input object can belong to all groups to a different extent and the sum of memberships has to be equal to 1.
- In fuzzy possibilistic methods it is assumed that each input object can belong to all groups to a different extent and the sum of memberships does not have to be equal to 1.

Clustering algorithms can display different, individual features. They can, for example, ignore "outlying"? training sets or generate groups with the shape different from spherical (as the most typical one). Due to the fact that in a data clustering task we do not have reference signals, the data clustering process is unsupervised learning. The quality of data clustering is evaluated using indicators that allow us to verify the similarity of data in clusters and the diversity of clusters. These indicators can also become a base of the mechanism for selecting the number of clusters and for example, the number of fuzzy rules [2, 12, 21, 30, 31, 35, 36] resulting from it.

The purpose of initialization implemented with the use of data clustering methods is spacing of input and output fuzzy sets taking into account, in particular, the areas in which there is the greatest density of these sets from the learning sequence used.

The following part of the section contains an example of parameters initialization of Gaussian fuzzy sets of the form (4.29) with the use of the fuzzy c-means (FCM) algorithm [6], which is a well-known method of data clustering.

The FCM algorithm uses a learning sequence for initialization. It contains Z learning sets with the structure described in Sect. 2.3.1 in the context of formula (2.18). On the basis of these sets, a matrix $\mathbf{X}$ of size $Z \times (n+m)$ is initialized. It stores input learning sets for the FCM algorithm:

$$\mathbf{X} = \begin{bmatrix} \bar{x}_{1,1} & \dots & \bar{x}_{n,1} & d_{1,1} & \dots & d_{m,1} \\ \vdots & & \vdots & \vdots & & \vdots \\ \bar{x}_{1,Z} & \dots & \bar{x}_{n,Z} & d_{1,Z} & \dots & d_{m,Z} \end{bmatrix}^T . \tag{5.1}$$

On the basis of this matrix, clusters centre matrix $\mathbf{V}$ of size $N \times (n+m)$ is initialized. The easiest method is the uniform or random initialization, although there are also more advanced techniques (see e.g. [50]). After the initialization of matrices $\mathbf{V}$ and $\mathbf{X}$, the FCM algorithm starts. Its operation takes a certain number of steps [6], and it results in adequately selected matrices $\mathbf{V}$ and $\mathbf{U}$ of size $N \times Z$. The elements of matrix $\mathbf{U}$ represent memberships of input sets in the clusters. The FCM algorithm performs limited fuzzy clustering, so for each fuzzy set the sum of its memberships is equal to 1. The elements of matrices $\mathbf{U}$ and $\mathbf{V}$ are used for spacing input and output fuzzy sets. Their centers can be spaced as follows:

$$\begin{cases} \bar{x}_{i,k}^{A} = v_{k,i} \\ \bar{x}_{j,k}^{B} = v_{k,n+j} . \end{cases} \tag{5.2}$$

Then, the initial values of the parameters responsible for the width of input and output fuzzy sets A_i^k and B_j^k can be determined as follows:

$$\begin{cases} \sigma_{i,k}^{A} = \sqrt{\dfrac{\sum\limits_{z=1}^{Z} u_{k,z}{}^{\theta} \cdot \left(\bar{x}_{z,i} - v_{k,i}\right)^2}{\sum\limits_{z=1}^{Z} u_{k,z}^{\theta}}} \\ \sigma_{j,k}^{B} = \sqrt{\dfrac{\sum\limits_{z=1}^{Z} u_{k,z}{}^{\theta} \cdot \left(d_{z,j} - v_{k,n+j}\right)^2}{\sum\limits_{z=1}^{Z} u_{k,z}^{\theta}}} , \end{cases} \tag{5.3}$$

where $\theta \geq 1$ is a parameter of the FCM algorithm, known as the fuzzification coefficient [6]. Similarly, the FCM algorithm can be used for initial spacing of discretization points.

Table 5.1 contains a set of parameters of the system (4.28) and typical methods of their initialization before starting gradient learning. However, the issue of fuzzy system initialization can be looked at more broadly:

Table 5.1 A list of system parameters (4.28) and typical initialization methods before gradient learning

Item no.	Parameter type	Initialization of constant values	Initialization of random values	Even initialization	Initialization using unsupervised learning
1.	Parameters of input A_i^k and output B_j^k fuzzy sets	Yes	Yes	Yes	Yes
2.	Weights of importance of input fuzzy sets $w_{i,k}^A$, output fuzzy sets $w_{j,k}^B$ and fuzzy rules w_k^{rule}	Yes	Yes	No	Yes
3.	Shape parameters of triangular parameterized norms p_k^{τ}, p_k^{inf} and p_k^{agr}	Yes	Yes	No	No
4.	Discretization points $\bar{y}_{j,r}^{\text{def}}$	Yes	Yes	Yes	Yes
5.	Inference-type parameters ν	Yes	Yes	No	No

- When initializing a single fuzzy system we can use a few different initialization methods (e.g. those given in Table 5.1).
- When initializing fuzzy systems based on data clustering techniques we can use different algorithms. Then, their individual features have impact on specific properties of the system being initialized.
- Initialization can also be performed on the basis of a pool of solutions (i.e. fuzzy systems), from which the best ones are selected (in the context of adopted criteria) for gradient learning. In this case the operation of a gradient algorithm is reduced to tuning a pre-selected solution. Using a pool can also be important in on-line applications.
- Learning and initialization can undergo multithreaded implementation and each thread can represent a separate algorithm. Algorithms working parallel do not have to be of the same type. In this case a specific type of population is obtained.
- In more complex initialization techniques of fuzzy systems the boundary between learning and initialization can be blurred. It occurs e.g. when initialization (as in population algorithms) has sufficiently effective mechanisms for the exploration

and exploitation of the consideration space. Then, it plays the role of a specific population algorithm.

The issue of fuzzy systems initialization before learning with the use of population-based algorithms is related to fuzzy systems initialization before gradient learning. It has been discussed in a number of interesting papers [3, 13, 16, 17, 20, 22, 26, 27, 29, 37, 39–42, 48, 52, 57].

5.2 Gradient Learning of Fuzzy Systems

General remarks on gradient algorithms used in fuzzy systems learning can be summarized as follows:

- They are well suited for selecting fuzzy system continuous parameters which have no limitations imposed on the range of their values. Thus, they can be used to learn e.g. fuzzy sets parameters, shape parameters of parameterized triangular norms as well as discretization points.
- They are well suited for selection of fuzzy system continuous parameters which have limitations imposed on the range of their values. Such parameters are, for example, weights of importance, shape parameters of some parameterized triangular norms (Tables 4.1 and 4.2) and the inference-type parameter. In this case, we have to use a suitable mechanism to prevent these parameters' value from exceeding the allowed range. A simple and well-proven solution consists in that the parameters learned with limitations are replaced by so-called range limitation functions [9, 45, 47]. Then, the values returned by these functions will take into account the imposed limitations and their arguments will be able to freely change their values in the entire real numbers domain. The range functions should be differentiable and continuous functions. In practice, their form can be chosen to allow them to implement different limitations imposed on the parameters. An example of a range function is a sigmoid:

$$f_z(x) = \frac{p_{z3}}{1 + \exp\left(-\left(p_{z1} \cdot x - p_{z2}\right)\right)} + p_{z4}, \tag{5.4}$$

  which for parameters $p_{z1} = 10$, $p_{z2} = 5$, $p_{z3} = 1$ and $p_{z4} = 0$ returns values from the range $(0, 1)$ for $x \in R$. A set of functions which can be range functions is presented in Table 5.2.
- It is difficult to use them for building the structure of a fuzzy system. Therefore, it is difficult to select the number of the system rules, the number of fuzzy sets, the type of operators (aggregation, inference and defuzzification), the type of membership functions used, the subset of key input attributes of the system, etc. For this reason, the following supporting methods are very important in interpretable fuzzy systems designing: initialization (Sect. 5.1), tuning (Sect. 5.3), merging, and reduction (Sects. 5.4 and 5.5) algorithms.

Table 5.2 A list of typical functions which can work as range functions

Item no.	Function name	Function formula
1.	Sigmoid type function with left-hand and right-hand limits	$f_z(x) = \frac{p_{z3}}{1+\exp(-(p_{z1} \cdot x - p_{z2}))} + p_{z4}$
2.	Hyperbolic tangent type function with left-hand and right-hand limits	$f_z(x) = p_{z3} \cdot \tanh(p_{z1} \cdot x - p_{z2}) + p_{z4}$
3.	Bell type function	$f_z(x) = \frac{p_{z3}}{1+\exp\left(-\left(p_{z1} \cdot \sqrt{x^2} - p_{z2}\right)\right)} + p_{z4}$
4.	Function with left-hand limits	$f_z(x) = \frac{p_{z3}}{1+\exp\left(-\left(p_{z1} \cdot \sqrt{x^2} - p_{z2}\right)\right)} + p_{z4}$
5.	Function with right-hand limits	$f_z(x) = \frac{x}{1+\exp(p_{z1} \cdot x + p_{z2})} - p_{z3}$

The following part of this chapter shows an example how to learn fuzzy systems using an algorithm belonging to the gradient algorithms group i.e. the backpropagation algorithm [38, 51, 53, 54]. It is a popular algorithm which was originally used for the learning of artificial neural networks with the teacher. Its purpose is to minimize some error measure Q on the fuzzy system output. It is a function of the input signals vector of the system $\bar{\mathbf{x}}$ and the parameters vector of the system $\mathbf{p} = [p_1, \ldots, p_P]^T$. It can be expressed e.g. by formula (2.18). In order to minimize it the algorithm, in each step, finds a proper direction vector $\mathbf{g} = [g_1, \ldots, g_P]^T$ for the parameters vector of system $\mathbf{p}$. Vector $\mathbf{g}$ is determined using approximation of error measure Q by Taylor series expansion in the vicinity of actual vector $\mathbf{p}$. To simplify it a linear extension is used. On the basis of the rule of the steepest descent we can write:

$$\mathbf{p} := \mathbf{p} - \eta \cdot \mathbf{g}(\mathbf{p}), \tag{5.5}$$

where $\eta \in [0, 1]$ is a so-called learning coefficient (the intensity of learning) and the vector of gradient $\mathbf{g}$ has the following form:

$$\mathbf{g}(\mathbf{p}) = \left[\frac{\partial Q}{\partial p_1}, \ldots, \frac{\partial Q}{\partial p_P}\right]^T = [\Delta p_1, \ldots, \Delta p_P]^T. \tag{5.6}$$

A detailed derivation of dependencies (5.5) and (5.6) can be found e.g. in [46, 51]. The way of updating parameters of the fuzzy system expressed by formula (5.5) can be used directly in the fuzzy system of the form (4.28) presented in Chap. 4, whose neural structure is shown in Fig. 5.1. Therefore, the updating formula of, for example, parameter p_1^{inf} takes the following form:

$$p_1^{\text{inf}} := p_1^{\text{inf}} - \eta \cdot \frac{\partial Q}{\partial p_1^{\text{inf}}} = p_1^{\text{inf}} - \eta \cdot \Delta p_1^{\text{inf}}, \tag{5.7}$$

where Δp_1^{inf} is the component of the gradient vector describing the correction value for parameter p_1^{inf}. Therefore, it is dependent on the used formula describing an error

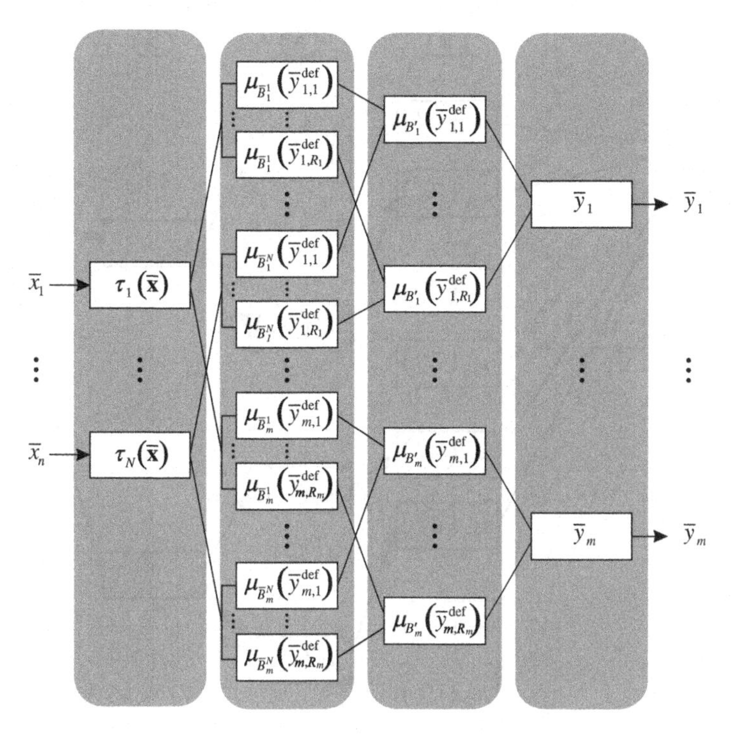

Fig. 5.1 The structure of the fuzzy system described by formula (4.28)

measure and the formula describing the fuzzy system output signal $\bar{y}_j$. The derivative function $\frac{\partial \bar{y}_i}{\partial p_1^{\mathrm{inf}}}$ is necessary for its determination. In the context of formula (4.28) it necessitates the determination of the following dependence:

$$\frac{\partial \bar{y}_i}{\partial p_1^{\mathrm{inf}}} = \frac{\partial \left(\dfrac{\sum\limits_{r=1}^{R_j} \bar{y}_{j,r}^{\mathrm{def}} \cdot \overset{N}{\underset{k=1}{\overset{\leftrightarrow}{\tilde{H}}^{*}}} \left(\begin{array}{c} \overset{\leftrightarrow}{\tilde{I}}^{*} \left(\begin{array}{c} \overset{n}{\underset{i=1}{\overset{\leftrightarrow}{\tilde{H}}^{*}}} \left(\begin{array}{c} \mu_{A_i^k}(\bar{x}_i); \\ w_{i,k}^{A}, p_k^{\tau}, 0 \end{array} \right), \mu_{B_j^k}\left(\bar{y}_{j,r}^{\mathrm{def}}\right); \\ 1, w_{j,k}^{B}, p_k^{\mathrm{inf}}, \nu \end{array} \right); \\ w_k^{\mathrm{rule}}, p_k^{\mathrm{agr}}, 1-\nu \end{array} \right)}{\sum\limits_{r=1}^{R_j} \overset{N}{\underset{k=1}{\overset{\leftrightarrow}{\tilde{H}}^{*}}} \left(\begin{array}{c} \overset{\leftrightarrow}{\tilde{I}}^{*} \left(\begin{array}{c} \overset{n}{\underset{i=1}{\overset{\leftrightarrow}{\tilde{H}}^{*}}} \left(\begin{array}{c} \mu_{A_i^k}(\bar{x}_i); \\ w_{i,k}^{A}, p_k^{\tau}, 0 \end{array} \right), \mu_{B_j^k}\left(\bar{y}_{j,r}^{\mathrm{def}}\right); \\ 1, w_{j,k}^{B}, p_k^{\mathrm{inf}}, \nu \end{array} \right); \\ w_k^{\mathrm{rule}}, p_k^{\mathrm{agr}}, 1-\nu \end{array} \right)} \right)}{\partial p_1^{\mathrm{inf}}}. \tag{5.8}$$

Determination of the derivative function of the form (5.8) (and derivative functions for the other system parameters) involves a considerable amount of work and an accompanying risk of making an error. Therefore, in our previous works we have proposed a simpler solution [47], which bases on a proper decomposition of the

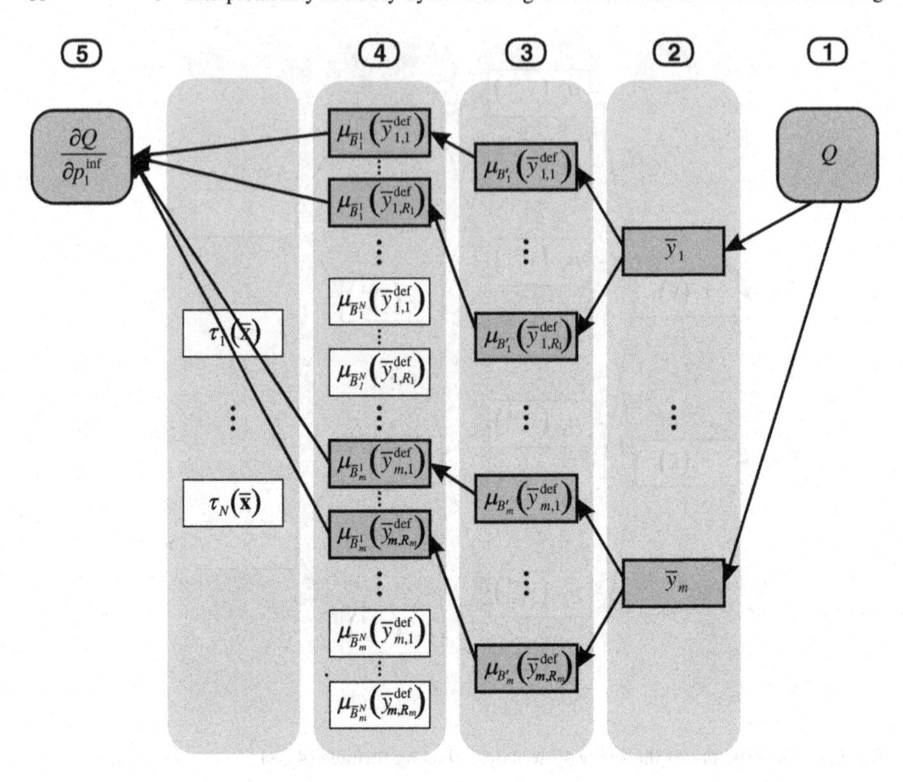

Fig. 5.2 The backpropagation in the method of the error backpropagation in the fuzzy system described by formula (4.28), shown on the basis of determination of values $\frac{\partial \bar{y}_i}{\partial p_1^{\text{inf}}}$

problem proposed in [45]. The idea of this decomposition is shown in Fig. 5.2 in the context of determination of correction $\frac{\partial \bar{y}_i}{\partial p_1^{\text{inf}}}$ for parameter p_1^{inf}. This figure shows a way of error backpropagation in the system (4.28), which involves the following steps:

- The first step consists in determining the current value of error measure and its propagation to the system outputs. For this purpose values of derivative functions $\frac{\partial Q}{\partial \bar{y}_1}, \ldots, \frac{\partial Q}{\partial \bar{y}_m}$ have to be determined. Error propagation to the output j consists in multiplying the determined error measure by the value of the derivative $\frac{\partial Q}{\partial \bar{y}_j}$.
- The second step consists in propagation of the current value of error measure by the defuzzification block described by formula (4.23). For this purpose values of derivative functions $\frac{\partial \bar{y}_1}{\partial \mu_{B'_1}\left(\bar{y}_{1,1}^{\text{def}}\right)}, \ldots, \frac{\partial \bar{y}_1}{\partial \mu_{B'_1}\left(\bar{y}_{1,R_1}^{\text{def}}\right)}, \ldots, \frac{\partial \bar{y}_m}{\partial \mu_{B'_m}\left(\bar{y}_{m,1}^{\text{def}}\right)}, \ldots, \frac{\partial \bar{y}_m}{\partial \mu_{B'_1}\left(\bar{y}_{m,R_m}^{\text{def}}\right)}$ have to be determined. Error propagation by the defuzzification block to the block $\mu_{B'_j}\left(\bar{y}_{j,r}^{\text{def}}\right)$ (especially on the input of the defuzzification block connected with

the output of the block $\mu_{B'_j}\left(\bar{y}^{\text{def}}_{j,r}\right))$ consists in multiplying the error measure propagated in the very last step by the value of the derivative $\frac{\partial \bar{y}_j}{\partial \mu_{B'_j}\left(\bar{y}^{\text{def}}_{j,r}\right)}$.

- The next steps (third and fourth) proceed in a similar way to the first two, in the direction shown in Fig. 5.2.
- The last step consists in determination of $\frac{\partial \bar{y}_i}{\partial p^{\text{inf}}_1}$. This is performed by summing the errors propagated by the blocks of the respective system (4.28) layers in the previous steps.

The presented way of determining $\frac{\partial \bar{y}_i}{\partial p^{\text{inf}}_1}$ eliminates the need for determining the derivative of the form (5.8). However, for each type of the block occurring in the system (4.28), it requires the determination of a full set of its derivatives. For example, the following derivatives have to be determined for the aggregation operator described by formula (4.25):

- The derivative with respect to the output $\mu_{\bar{B}^k_j}\left(\bar{y}^{\text{def}}_{j,r}\right)$, i.e. $\frac{\partial \mu_{B'_j}\left(\bar{y}^{\text{def}}_{j,r}\right)}{\partial \mu_{\bar{B}^k_j}\left(\bar{y}^{\text{def}}_{j,r}\right)}$.
- The derivative with respect to the weight w^{rule}_k, i.e. $\frac{\partial \mu_{B'_j}\left(\bar{y}^{\text{def}}_{j,r}\right)}{\partial w^{\text{rule}}_k}$.
- The derivative with respect to the parameter p^{agr}_k, i.e. $\frac{\partial \mu_{B'_j}\left(\bar{y}^{\text{def}}_{j,r}\right)}{\partial p^{\text{agr}}_k}$.
- The derivative with respect to the parameter ν, i.e. $\frac{\partial \mu_{B'_j}\left(\bar{y}^{\text{def}}_{j,r}\right)}{\partial \nu}$.

Table 5.3 contains a sample set of derivatives for the switchable Dombi h-function with weights of arguments. It can be an aggregation operator in the 1st and the 3rd layer of the system (4.28), whose neural structure is shown in Fig. 5.1.

The described decomposition of the learning algorithm notation greatly simplifies the problem of its selection for the specified type of a fuzzy system. It can be assumed that such approach represents a specific interpretability at the learning algorithm level. Thus, it is consistent with the assumptions of interpretable fuzzy system designing. A detailed notation of the error backpropagation algorithm for different varieties of the fuzzy system can be found in our previous works (e.g. in [10, 11, 47]).

5.3 Fuzzy System Tuning Aimed at Increasing Accuracy

In the rule of the steepest descent expressed by formula (5.5) a gradient vector is determined independently for each set from the learning sequence used. Next, on the basis of this vector all learnt parameters of a fuzzy system are corrected. It is realized Z times during one epoch. Thus, often in the last epoch of the learning process, the system parameters in the last iteration have an unfavorable value in the context of minimization of the system evaluation function. This function may be e.g. an error measure. Therefore, the aim of fuzzy system tuning is to find the direction vector

Table 5.3 A list of dependencies of error ε propagation (delivered to the output of the operator) by the switchable Dombi h-function with weights of arguments on its input (ε_{a_i}), weight (ε_{w_i}), shape parameter (ε_p) and compromise parameter (ε_ν)

Item no.	Notation of the function and its derivatives
1.	$\overset{\leftrightarrow}{H}^{*}_{D}(\mathbf{a};\mathbf{w},p,\nu) = \mathrm{neg}_Z\left(\widetilde{\mathrm{neg}}\left(\left(\left(1+\left(\sum_{i=1}^{n}\left(\widetilde{\mathrm{neg}}\left(\widetilde{\arg}(a_i;w_i,\nu);f_z(\nu)\right)^{-1}-1\right)^{f_{z1}(p)}\right)^{\frac{1}{f_{z1}(p)}}\right)^{-1}\right);f_z(\nu)\right)\right)$ $= 1-\widetilde{\mathrm{neg}}\left(\overset{\leftrightarrow}{h}^{*}_{D}(\mathbf{a};\mathbf{w},p,\nu);f_z(\nu)\right)$
2.	$\varepsilon_{a_i} = \left((2f_z(\nu)-1)^2\frac{\left(\overset{\leftrightarrow}{h}^{*}_{D}(\mathbf{a};\mathbf{w},p,\nu)^{-1}-1\right)^{1-f_{z1}(p)}}{\overset{\leftrightarrow}{h}^{*}_{D}(\mathbf{a};\mathbf{w},p,\nu)^{-2}}\cdot\frac{\left(\widetilde{\mathrm{neg}}(\widetilde{\arg}(a_i;w_i,\nu);f_z(\nu))^{-1}-1\right)^{-f_{z1}(p)}}{\left(1-\widetilde{\mathrm{neg}}(\widetilde{\arg}(a_i;w_i,\nu);f_z(\nu))\right)\widetilde{\mathrm{neg}}(\widetilde{\arg}(a_i;w_i,\nu);f_z(\nu))}\frac{\partial\widetilde{\arg}(a_i;w_i,\nu)}{\partial a_i}\right)\varepsilon$
3.	$\varepsilon_{w_i} = \left((2f_z(\nu)-1)^2\frac{\left(\overset{\leftrightarrow}{h}^{*}_{D}(\mathbf{a};\mathbf{w},p,\nu)^{-1}-1\right)^{1-f_{z1}(p)}}{\overset{\leftrightarrow}{h}^{*}_{D}(\mathbf{a};\mathbf{w},p,\nu)^{-2}}\cdot\frac{\left(\widetilde{\mathrm{neg}}(\widetilde{\arg}(a_i;w_i,\nu);f_z(\nu))^{-1}-1\right)^{-f_{z1}(p)}}{\left(1-\widetilde{\mathrm{neg}}(\widetilde{\arg}(a_i;w_i,\nu);f_z(\nu))\right)\widetilde{\mathrm{neg}}(\widetilde{\arg}(a_i;w_i,\nu);f_z(\nu))}\frac{\partial\widetilde{\arg}(a_i;w_i,\nu)}{\partial w_i}\right)\varepsilon$
4.	$\varepsilon_p = \frac{2f_z(\nu)-1}{f_{z1}(p)}\frac{\left(\overset{\leftrightarrow}{h}^{*}_{D}(\mathbf{a};\mathbf{w},p,\nu)^{-1}-1\right)^{1-f_{z1}(p)}}{\overset{\leftrightarrow}{h}^{*}_{D}(\mathbf{a};\mathbf{w},p,\nu)^{-2}}\left(\sum_{i=1}^{n}\frac{-\ln\left(\widetilde{\mathrm{neg}}(\widetilde{\arg}(a_i;w_i,\nu);f_z(\nu))^{-1}-1\right)}{\left(\widetilde{\mathrm{neg}}(\widetilde{\arg}(a_i;w_i,\nu);f_z(\nu))^{-1}-1\right)^{f_{z1}(p)}}+ -\frac{\ln\left(\left(\overset{\leftrightarrow}{h}^{*}_{D}(\mathbf{a};\mathbf{w},p,\nu)^{-1}-1\right)\right)}{\left(\overset{\leftrightarrow}{h}^{*}_{D}(\mathbf{a};\mathbf{w},p,\nu)^{-1}-1\right)^{-f_{z1}(p)}}\right)\frac{\partial f_{z1}(p)}{\partial p}\varepsilon$
5.	$\varepsilon_\nu = \overset{\leftrightarrow}{h}^{*}_{D}(\mathbf{a};\mathbf{w},p,\nu)\left(\left(\left(\overset{\leftrightarrow}{h}^{*}_{D}(\mathbf{a};\mathbf{w},p,\nu)^{-1}-1\right)^2-1\right)\frac{\partial f_z(\nu)}{\partial\nu}+ +\frac{(2f_z(\nu)-1)}{\left(\overset{\leftrightarrow}{h}^{*}_{D}(\mathbf{a};\mathbf{w},p,\nu)^{-1}-1\right)^{f_{z1}(p)-1}}\cdot\left(\sum_{i=1}^{n}\frac{\left(\widetilde{\mathrm{neg}}(\widetilde{\arg}(a_i;w_i,\nu);f_z(\nu))^{-1}-1\right)^{-f_{z1}(p)}}{\left(\widetilde{\mathrm{neg}}\left(\widetilde{\arg}(a_i;w_i,\nu);f_z(\nu)\right)-1\right)\cdot\widetilde{\mathrm{neg}}\left(\widetilde{\arg}(a_i;w_i,\nu);f_z(\nu)\right)}\right)\cdot\left((2f_z(\nu)-1)\frac{\partial\widetilde{\arg}(a_i;w_i,\nu)}{\partial\nu}+ +\left(2\widetilde{\arg}(a_i;w_i,\nu)-1\right)\frac{\partial f_z(\nu)}{\partial\nu}\right)\right)\varepsilon$

for which modification of the system parameters will have the most beneficial effect on system accuracy. However, a higher accuracy of the system may necessitate (as previously mentioned) using a smaller fuzzy rules base.

A sample realization of the final tuning algorithm (FTA) of a fuzzy system aimed at increasing its interpretability is shown in Fig. 5.3 [9]. It works as follows. First, the accuracy of the system which has been learnt (e.g. using the error backpropagation algorithm or another) and the system state are stored. A system state means a complete set of its current parameters. Next, for each learning set a learning process taking one iteration is performed, during which a new gradient vector is determined. After each iteration the system accuracy is determined. It is compared to the accuracy stored

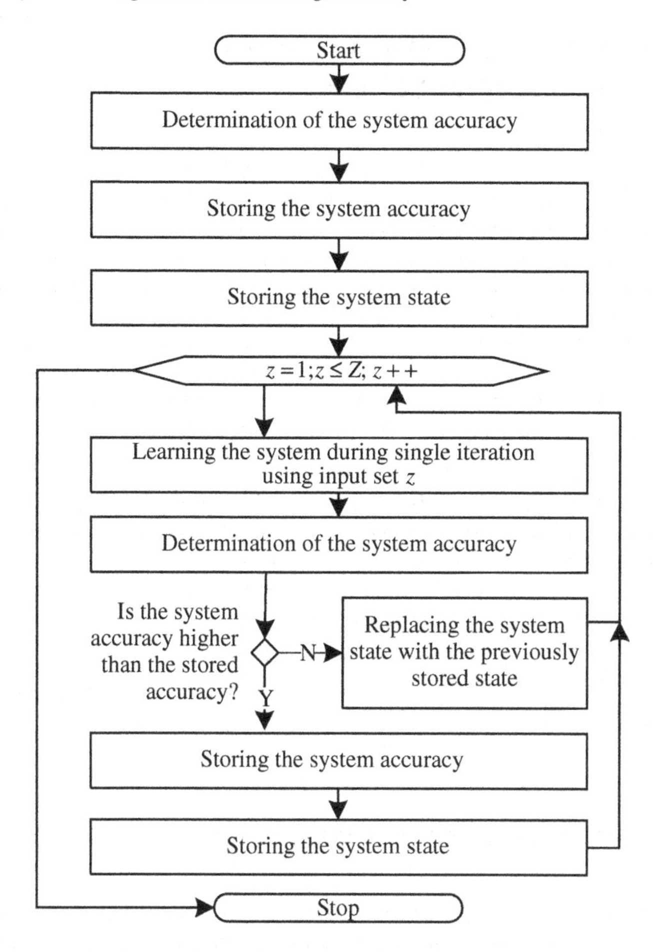

Fig. 5.3 Exemplary implementation of the algorithm of fuzzy system final tuning aimed at increasing accuracy

earlier. If the current system works more accurately than the system stored earlier, then the current state of the system and its accuracy are stored. Otherwise, they are replaced by the data stored earlier. An entire learning epoch is performed by analyzing successive input sets from the learning sequence in this way. The proposed procedure can be called a gradient procedure for exploitation of a given space, because the best solution is searched close to the solutions found earlier.

The final tuning algorithm can be used at the end of gradient learning. It is also an important component of other procedures, e.g. reduction procedures discussed in the following sections. In this case the algorithm repairs a reduced fuzzy system.

5.4 Fuzzy System Reduction Aimed at Maximum Simplification of the Structure

As already mentioned, gradient algorithms show little potential in terms of positive impact on fuzzy systems interpretability (Sect. 5.2). Therefore, researchers are looking for supporting solutions which would allow them to select e.g. the number of fuzzy system rules, the number of fuzzy sets and the subset of the key input attributes. Such solution can be provided by a reduction of fuzzy systems that have been learnt e.g. using a gradient algorithm. Reduction mechanism can be mainly used in two ways:

- The first method relies on maximization of the fuzzy system reduction degree without compromising its accuracy (or with a "controlled compromise"). Therefore, it is associated with removing as many components of a reduced fuzzy system as possible, i.e. rules, input attributes, fuzzy sets and discretization points. Sample algorithms representing this type of reduction mechanism are presented in Sects. 5.4.1 and 5.4.2.
- The second method relies on reduction aimed at increasing accuracy. Therefore, it is based on appropriate searching and removing of system components, i.e. rules, input fuzzy sets attributes and discretization points, which adversely affect accuracy and interpretability. Section 5.5 presents algorithms representing this type of reduction mechanism.

5.4.1 Method of Consecutive Eliminations

The consecutive eliminations algorithm (CEA) is a method which uses reduction mechanism in order to achieve the maximum simplification of a fuzzy system structure while maintaining acceptable working accuracy. This method consists in consecutive elimination of those system components which are conflicting, inactive, and inert to the operation of the system. These elements are searched consecutively within the groups of fuzzy rules, input attributes, fuzzy sets and discretization points. The reduction order has been set to start with these system components which have the greatest impact on its rule base interpretability.

The main block scheme of the CEA algorithm is shown in Fig. 5.4. It starts its operation with determining the initial accuracy of a reduced system treated as the accuracy of a reference system. The reference system is a fuzzy system which has been learnt e.g. using a gradient algorithm. The determined accuracy of a reference system is stored as in the case of the system state. The system state is represented by a set of its parameters. After storing the reference system state and its parameters, the main reduction process starts. First, the system fuzzy rules are reduced (Fig. 5.5). Next, there are attempts made to delete the rules whose weights of importance are equal to or less than a certain threshold value $w_{red} \in [0, 1]$. After removing each of these rules the system is repaired using the final tuning algorithm discussed in

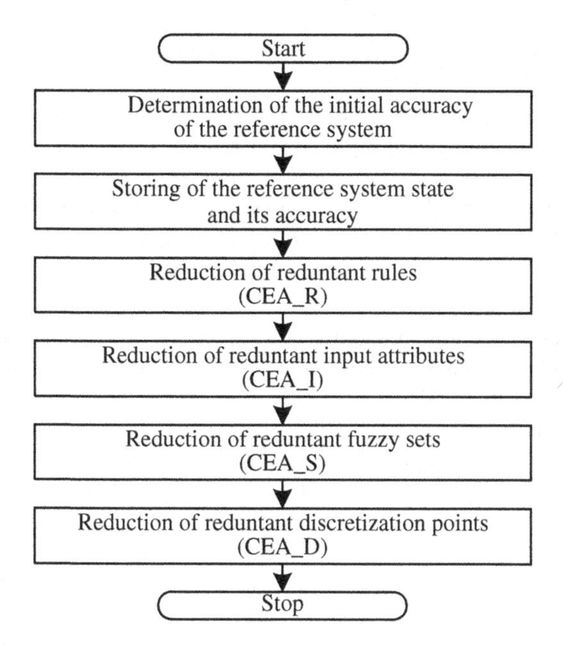

Fig. 5.4 Consecutive eliminations algorithm - the main procedure

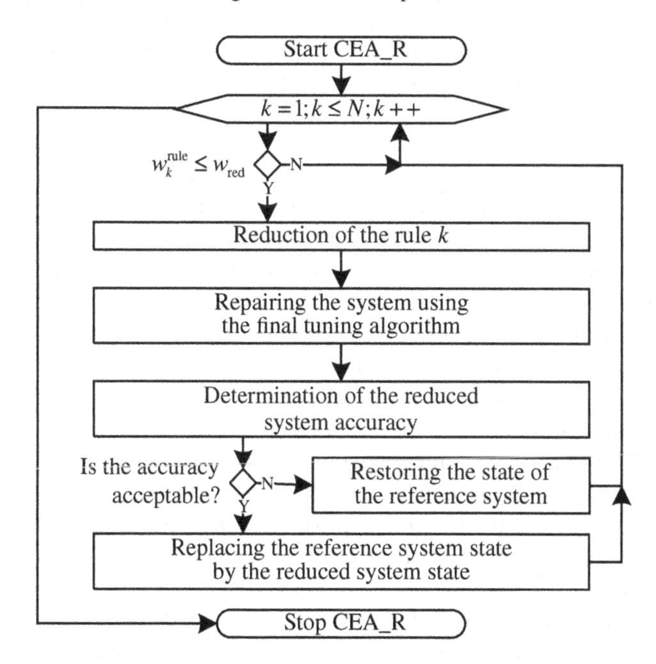

Fig. 5.5 Consecutive eliminations algorithm - the rules reduction procedure

Sect. 5.3. Then, the impact of the reduction on the system accuracy is checked. If it is acceptable (i.e. for example the system accuracy after the conducted reduction is not worse than in the case of the reference system), then the reduction is approved. Approval of the reduction means replacing the reference system state by the reduced system state. If the reduction compromises the system performance and it is not acceptable, then the state of the reduced system is replaced by the state of the reference system. Following this procedure all the fuzzy rules of the system are consecutively analyzed. As a result, all the rules which have not had a beneficial impact on the system accuracy are deleted.

After the rules' reduction, input attributes, fuzzy sets and discretization points are consecutively deleted. Since block diagrams of these procedures are analogous to the scheme presented in Fig. 5.5, they are not presented here. It is worth mentioning that in the discretization points reduction procedure there are no weights, therefore all the points are successively analyzed. Yet, in the input attributes reduction procedure their weights can be taken into account. These weights can be determined as follows:

$$w_i^{\mathrm{inp}} = \frac{1}{N} \cdot \sum_{k=1}^{N} w_{i,k}^{A}. \tag{5.9}$$

Further comments concerning the CEA algorithm can be summarized as follows:

- In the CEA algorithm weights control can be removed as in the procedure of discretization points reduction. A similar effect can be achieved by setting $w_{\mathrm{red}} = 1$.
- The CEA algorithm can be focused on maximization of accuracy. Then, the condition of the reduction implementation can be e.g. achieving a greater accuracy than the reference system accuracy previously stored.
- The CEA algorithm can take into account different interpretability criteria.
- The CEA algorithm does not perform an initial analysis of the whole set of fuzzy system components in order to select the one whose reduction has the most beneficial impact on e.g. accuracy. Therefore, it belongs to the group of algorithms aimed at maximization of the reduction level.
- The literature presents a number of attempts made to select components of a fuzzy system for reduction (e.g. [7, 18, 19, 23–25, 28, 34, 43, 44, 49]). However, it is a complex issue in which the reduction of theoretically selected fuzzy system components may not be acceptable in terms of used evaluation criteria. The CEA algorithm can cooperate with other approaches, but its advantage is that it allows us to perform an immediate verification of the effects of a reduction carried out sequentially.
- After it has been slightly modified the CEA algorithm can allow us to determine the importance of each component of a fuzzy system in the context of an adopted evaluation criterion.
- Having been slightly modified the CEA algorithm can generate as a function of complexity. The complexity can be expressed e.g. by the number of fuzzy rules, the

number of input attributes, the number of fuzzy sets, the number of discretization points, and the number of system parameters which are learnt.

- The CEA algorithm should be treated as a sample method which implements a basic strategy for the selection of elements which should be reduced and the acceptance of reduction carried out only on the basis of accuracy.

5.4.2 *Method of Consecutive Merging*

The approach described in Sect. 5.4.1 (and algorithm CEA representing it) is aimed at reduction of those system components which have not had a negative impact on system operation accuracy. However, reduction can also be implemented by merging those system components which work in a similar way. This process bases on different similarity measures [4, 5, 14, 32, 58]. Merging contributes to increasing the coherence of fuzzy system components. This coherence is beneficial in the context of interpretable fuzzy system designing.

Merging of similar fuzzy sets is performed most frequently. Then, they can be shared between a few rules. Merging should be implemented within the scope of the same input attributes or outputs. However, it can be performed only if it does not compromise system accuracy. An example of the algorithm using a reduction mechanism by merging is the consecutive merging algorithm (CMA). The final effect of merging is a set of fuzzy sets with minimized cardinality.

The main block scheme of the CEA algorithm is shown in Fig. 5.6 [9]. First, the accuracy of its reference system is determined. Next, the state of the reference system and its accuracy are stored. Then, the main merging procedure starts. Input fuzzy sets and next output fuzzy sets are analyzed. For each combination of input fuzzy sets associated with a common input attribute the value of their similarity measure is determined. If it is greater than the assumed threshold value of the fuzzy

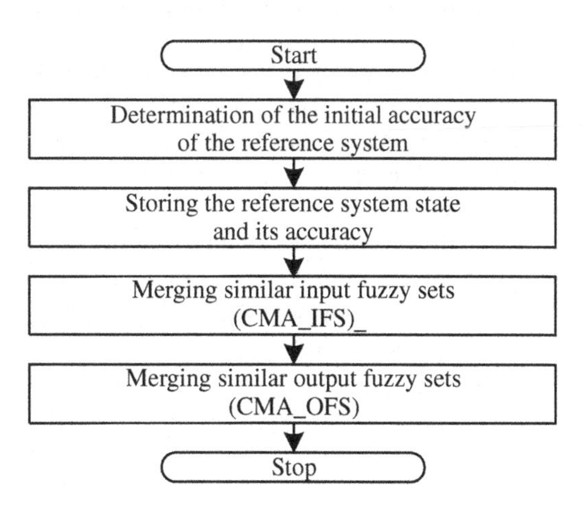

Fig. 5.6 Consecutive merging algorithm - the main procedure

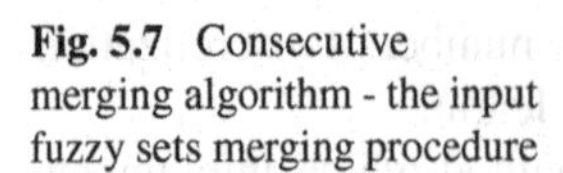
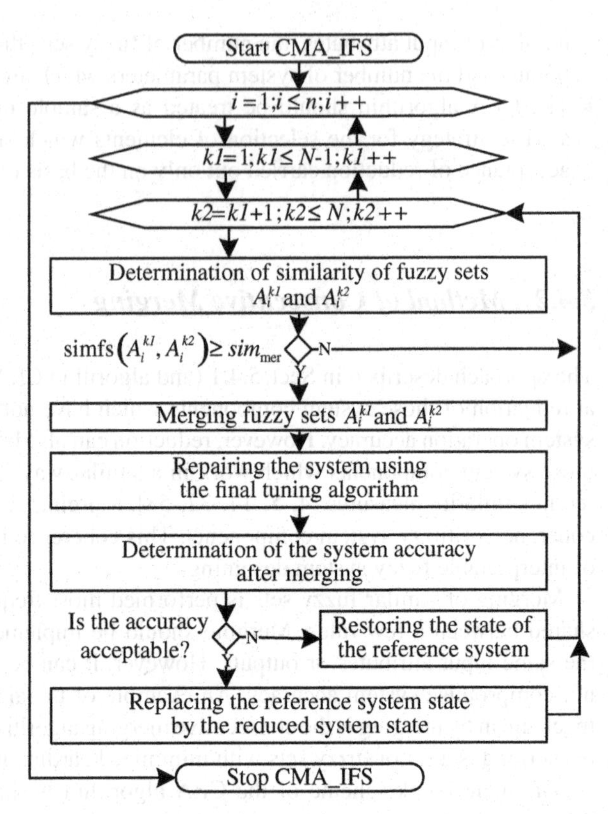

Fig. 5.7 Consecutive merging algorithm - the input fuzzy sets merging procedure

sets similarity $sim_{\text{mer}} \in [0, 1]$, then, these sets are merged. After the merging, the system is repaired with the use of the final tuning algorithm presented in Sect. 5.3. Moreover, the accuracy of the system which has just undergone the merging process is determined. If the accuracy is found to be acceptable, then the merging is approved. It means replacing the state of the reference system by the state of the system in which the merging algorithm was operated. If the obtained accuracy is not satisfactory, then, the merging is reversed. In this case the state of the reduced system is replaced by the state of the reference system (Fig. 5.7). After an analysis of the fuzzy sets associated with all the input attributes, a merging procedure in the group of output fuzzy sets is performed. Since it is implemented using an analogous procedure, its block scheme is not presented in here.

The CMA algorithm bases on a properly determined similarity measure of fuzzy sets [4, 5, 14, 32, 58]. A popular fuzzy set similarity measure is the Jaccard index, which is defined as the quotient of the power of the joint part of fuzzy sets and the power of the sum of these sets:

$$\text{simfs}\left(A_i^{k1}, A_i^{k2}\right) = \frac{\left|A_i^{k1} \cap A_i^{k2}\right|}{\left|A_i^{k1} \cup A_i^{k2}\right|}. \tag{5.10}$$

The Jaccard index has values from the range [0, 1]. Values close to 0 indicate that compared fuzzy sets are different and values close to 1 mean that such sets are similar. A discrete form of the Jaccard index (5.10) for fuzzy sets expressed by the Gaussian membership function (4.29) can be defined as follows:

$$\operatorname{simfs}\left(A_i^{k1}, A_i^{k2}\right) = \frac{\sum_{j=0}^{J-1} \min\left\{\begin{array}{l} \mu_G\left(\bar{x}_{\min} + j \cdot \frac{\bar{x}_{\max}-\bar{x}_{\min}}{J-1}; \bar{x}_{i,k1}^A, \sigma_{i,k1}^A\right), \\ \mu_G\left(\bar{x}_{\min} + j \cdot \frac{\bar{x}_{\max}-\bar{x}_{\min}}{J-1}; \bar{x}_{i,k2}^A, \sigma_{i,k2}^A\right) \end{array}\right\}}{\sum_{j=0}^{J-1} \max\left\{\begin{array}{l} \mu_G\left(\bar{x}_{\min} + j \cdot \frac{\bar{x}_{\max}-\bar{x}_{\min}}{J-1}; \bar{x}_{i,k1}^A, \sigma_{i,k1}^A\right), \\ \mu_G\left(\bar{x}_{\min} + j \cdot \frac{\bar{x}_{\max}-\bar{x}_{\min}}{J-1}; \bar{x}_{i,k2}^A, \sigma_{i,k2}^A\right) \end{array}\right\}}, \tag{5.11}$$

where J is the number of comparison points of fuzzy sets A_i^{k1} and A_i^{k2}, $\bar{x}_{i,k1}^A$ and $\bar{x}_{i,k2}^A$ are the centers of compared fuzzy sets, $\sigma_{i,k1}^A$ and $\sigma_{i,k2}^A$ are the widths of compared fuzzy sets, while $\bar{x}_{\min}$ and $\bar{x}_{\max}$ are determined on the basis of the following dependency:

$$\left\{\begin{array}{l} \bar{x}_{\min} = \min\left\{\bar{x}_{\min 1}, \bar{x}_{\min 2}\right\} = \min\left\{\begin{array}{l} \bar{x}_{i,k1}^A - \sigma_{i,k1}^A \cdot \sqrt{-\ln(\psi)}, \\ \bar{x}_{i,k2}^A - \sigma_{i,k2}^A \cdot \sqrt{-\ln(\psi)} \end{array}\right\} \\ \bar{x}_{\max} = \max\left\{\bar{x}_{\max 1}, \bar{x}_{\max 2}\right\} = \max\left\{\begin{array}{l} \bar{x}_{i,k1}^A + \sigma_{i,k1}^A \cdot \sqrt{-\ln(\psi)}, \\ \bar{x}_{i,k2}^A + \sigma_{i,k2}^A \cdot \sqrt{-\ln(\psi)} \end{array}\right\}. \end{array}\right. \tag{5.12}$$

The interpretation of constant $\psi \in (0, 1)$ in formula (5.12) is shown in Fig. 5.8. Its value should be close to 0. The values of $\bar{x}_{\min 1}$, $\bar{x}_{\min 2}$, $\bar{x}_{\max 1}$ and $\bar{x}_{\max 2}$ are computed using the equations of the form:

$$\left\{\begin{array}{l} \mu_G\left(\bar{x}; \bar{x}_{i,k1}^A, \sigma_{i,k1}^A\right) = \psi \Rightarrow \left\{\bar{x}_{\min 1}, \bar{x}_{\max 1}\right\} \\ \mu_G\left(\bar{x}; \bar{x}_{i,k2}^A, \sigma_{i,k2}^A\right) = \psi \Rightarrow \left\{\bar{x}_{\min 2}, \bar{x}_{\max 2}\right\}. \end{array}\right. \tag{5.13}$$

The presented way of determining fuzzy sets similarity refers to the sets described by the Gaussian membership functions of the form (5.8), whose values tend asymptotically to 0. In the case of using other membership functions, e.g. triangular functions, the comparison range of fuzzy sets $[\bar{x}_{\min}, \bar{x}_{\max}]$ is indicated without the need to set the constant ψ.

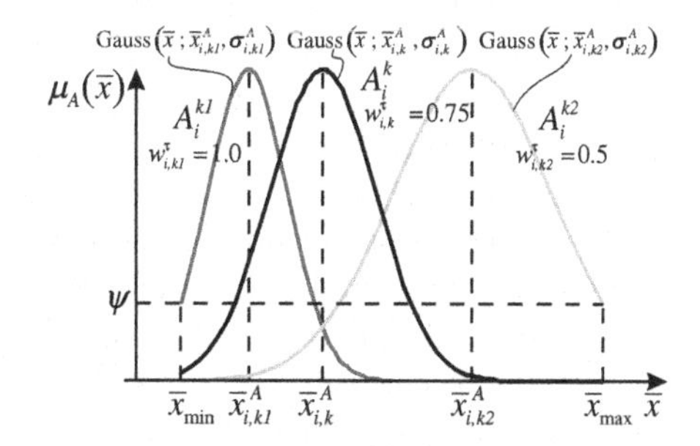

Fig. 5.8 Sample illustration of a merging procedure of two fuzzy sets with weights of arguments expressed by the Gaussian membership function (4.29)

The parameters of a fuzzy set obtained by merging two fuzzy sets expressed by the Gaussian membership function (4.29) and having weights of arguments are determined as follows:

$$\begin{cases} \bar{x}_{i,k}^{A} = \frac{\bar{x}_{i,k1}^{A} \cdot w_{i,k1}^{A} + \bar{x}_{i,k2}^{A} \cdot w_{i,k2}^{A}}{w_{i,k1}^{A} + w_{i,k2}^{A}} \\ \sigma_{i,k}^{A} = \frac{\sigma_{i,k1}^{A} \cdot w_{i,k1}^{A} + \sigma_{i,k_2}^{A} \cdot w_{i,k2}^{A}}{w_{i,k1}^{A} + w_{i,k2}^{A}} \\ w_{i,k}^{A} = \frac{w_{i,k1}^{A} + w_{i,k2}^{A}}{2}. \end{cases} \tag{5.14}$$

Additional comments on the CMA algorithm can be summarized as follows:

- In the CMA algorithm similarity control does not need to be applied. A similar effect can be achieved by setting $sim_{\text{mer}} = 0$.
- The CMA algorithm can be aimed at achieving maximization of system accuracy. Then, the approval condition for a merging procedure could be e.g. achieving accuracy greater than that of the reference system used.
- The CMA algorithm does not perform an initial analysis of all fuzzy sets combinations in order to select the combination on which merging is going to have the most beneficial impact, e.g. accuracy. Therefore, it belongs to the group of algorithms aimed at maximization of the reduction level.
- The CMA algorithm can be extended with a block which implements merging of other system components, e.g. fuzzy rules or discretization points. In such case it is advisable to use (or develop) a proper similarity measure.
- The CMA algorithm should be treated as a sample method from the group of methods implementing a basic strategy for selection of elements for merging and acceptance of merging on the basis of achieved accuracy.

5.5 Fuzzy Systems Reduction Aimed at Structure Simplification and Increase of Accuracy

The methods for fuzzy system reduction considered in Sect. 5.4 are aimed at the maximum structure simplification, especially in the context of the fuzzy rules base. In this case reduction is a mechanism increasing their interpretability. In this section we discuss an approach using reduction of those system elements whose removal is the most advantageous in terms of accuracy. The algorithms presented in Sects. 5.5.1 and 5.5.2 differ from each other in terms of the selection strategy of elements which are going to be reduced. The algorithm described in Sect. 5.5.1 searches for them locally, i.e. in a group of fuzzy rules, input attributes, fuzzy sets and discretization points, while the algorithm described in Sect. 5.5.2 looks for elements which are going to be reduced globally, i.e. in a group of all system components. Using reduction and merging mechanisms (i.e. reduction by merging) can support not only gradient algorithms but also population methods.

5.5.1 Method of the Most Favorable Local Eliminations

The most favorable local eliminations algorithm (MFLEA) searches within each group of parameters (i.e. rules, input attributes, fuzzy sets, discretization points) for the element whose reduction is the most favorable from the point of view of system accuracy. The name of the algorithm is based on its mode of operation. Its main block scheme is consistent with the scheme of the CEA algorithm presented in Fig. 5.4. However, the way in which component blocks are implemented is different. In the context of the fuzzy rules reduction it is conducted in accordance with the scheme shown in Fig. 5.9. First, the initial accuracy of the reference system is determined. The accuracy and state of the reference system are stored. Then, the main reduction

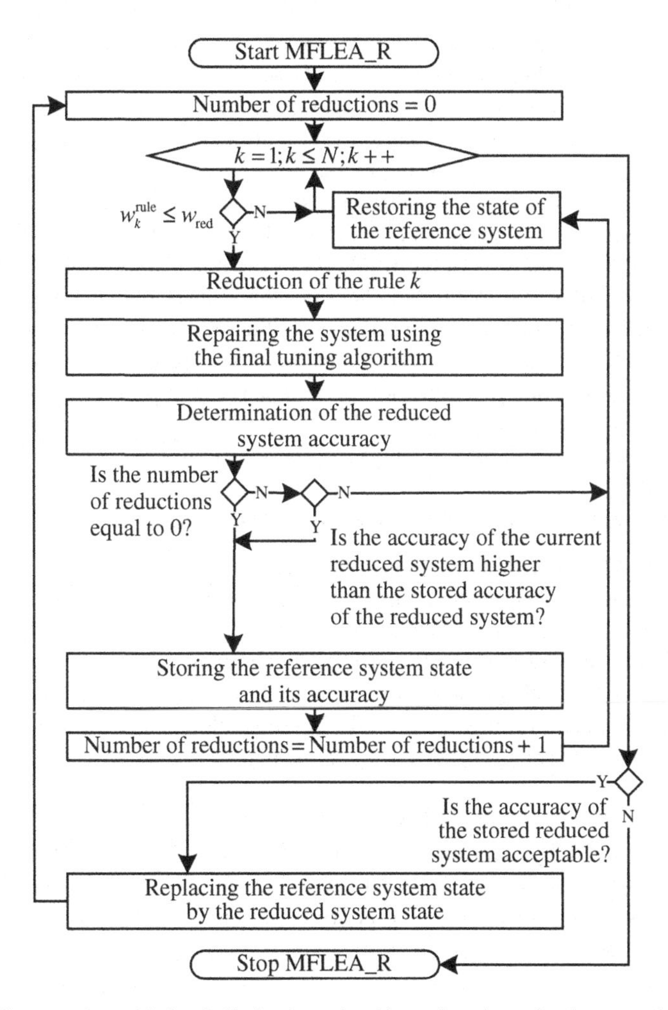

Fig. 5.9 The most favorable local eliminations algorithm - the rules reduction procedure

process starts. At the beginning the reduction counter value is set to 0. Next, the rules whose weights of importance are equal or less than a certain threshold value w_{red} are removed from the system. Each removal of a rule is followed by the system being repaired with the use of the final tuning algorithm discussed in Sect. 5.3. Moreover, the accuracy of the reduced system is determined. If the reduction counter indicates 0, the state of the reduced system and its accuracy are stored (this is the first step of the algorithm). If the value of the reduction counter is greater than zero, it means that the reduced system has been already stored. In this case it is checked whether the accuracy of the current reduced system is greater than the accuracy of the reduced system stored earlier. If it is, the state of the current reduced system and its accuracy are stored. In this way, the previous information on the system is lost. If it turns out that the current reduced system does not work better than the reduced system (stored in the previous steps of the algorithm), then the information on its state is not stored. Before attempting the reduction of the next rule, the reference system state is restored. A sequentially realized reduction of each of all fuzzy system rules finally results in the state of the reduced system whose accuracy following such reduction is the highest. Then, the algorithm goes to the conditional block in which it is decided whether to approve the reduction. In practice the approval of the reduction consists in replacing the reference system state by the stored state of the reduced system. This decision is based on a comparison of the achieved accuracy. If the reduction of a rule is approved, then an analysis of the whole set of the other rules is performed again. This makes it possible to search for another rule which can be reduced (Fig. 5.9). If the condition concerning an acceptable accuracy of the reduced system is not satisfied, then the reduction in the set of elements of the next type (i.e. in the sets of input attributes, fuzzy sets and discretization points) starts. The process performs in the same way as in the case of fuzzy rules. Because in the case of discretization points their weights are not available, an analysis of these weights cannot be performed and has to be left out.

Additional remarks on the MFLEA algorithm can be summarized as follows:

- In the MFLEA algorithms weights control can be removed, similar to the CEA algorithm. Due to this all the components of a fuzzy system can be reduced. A similar effect can be achieved setting $w_{\text{red}} = 1$.
- The MFLEA algorithm can take into account various criteria of fuzzy systems evaluation, not only the accuracy criterion.
- Following a small modification the MFLEA algorithm can allow us to determine the importance of each fuzzy system component in the context of an used evaluation criterion.
- The MFLEA algorithm might have its equivalent in the group of algorithms targeted at reduction by merging (the method of the most favorable local merging).
- The MFLEA algorithm should be treated as a sample method from the group of methods using reduction aimed at structure simplification and increasing of accuracy.

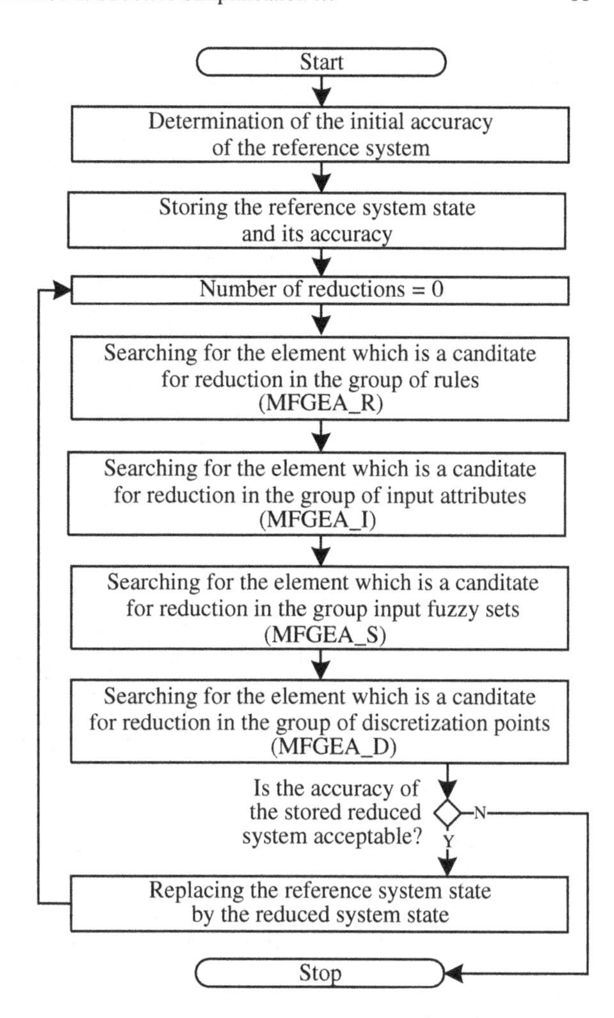

Fig. 5.10 The most favorable global eliminations algorithm - the main procedure

5.5.2 *Method of the Most Favorable Global Eliminations*

The most favorable global eliminations algorithm (MFGEA) works in a similar way to the MFLEA algorithm discussed in Sect. 5.5.1. The main difference lies in the fact that the MFGEA searches for the elements for reduction not in the local groups of the system components but in all the system components (globally). The name of the algorithm is based on it. Its main block scheme is presented in Fig. 5.10.

The MFGEA algorithm starts with determination of the initial accuracy of its reference system. It is stored together with the system state. Moreover, the value of the reduction counter is set to zero. Next, a search for the component which will be a candidate for reduction is performed. This component is searched for among all the elements supposed to undergo the reduction process: rules, input attributes,

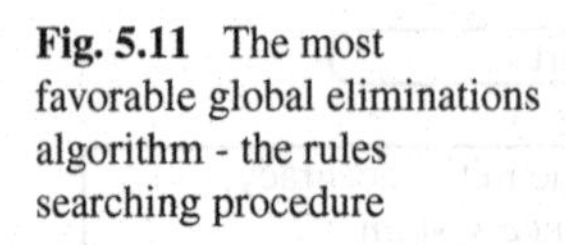

Fig. 5.11 The most favorable global eliminations algorithm - the rules searching procedure

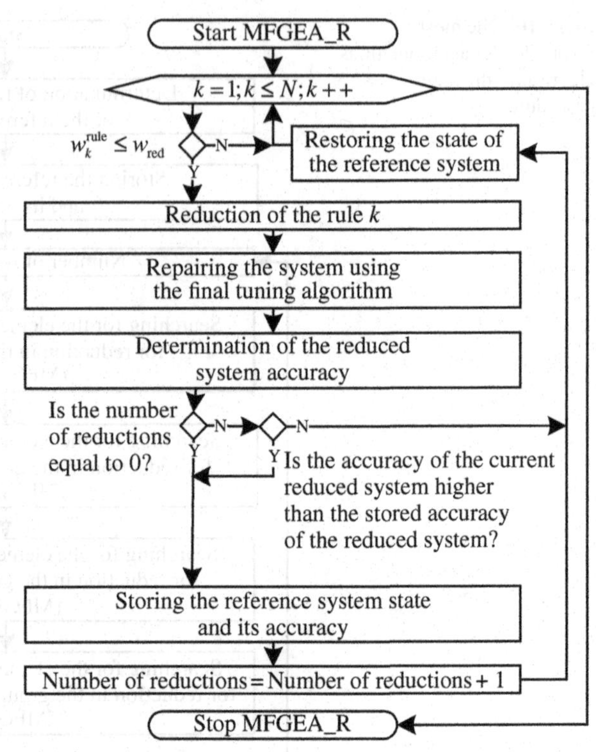

fuzzy sets and discretization points. Figure 5.11 shows a way of searching for the element for the reduction in a fuzzy rules group. In the other groups the search is implemented in a similar way. The management mechanism of the reference system state, reduced system state and reduction counter is the same like in the MFLEA algorithm.

Additional comments on the MFGEA algorithm can be summarized as follows:

- In the MFGEA algorithm weights control can be removed, similar as in the CEA and MFLEA algorithms. Due to this all the components of a fuzzy system can be reduced. A similar effect can be achieved setting $w_{red} = 1$.
- The MFGEA algorithm can take into account various criteria of fuzzy system evaluation, not only the accuracy criterion.
- The MFGEA algorithm following a small modification can allow us to determine the importance of each fuzzy system component in the context of the used evaluation criterion.
- The MFGEA algorithm might have its equivalent in the group of algorithms aimed at reduction by merging (the method of the most favorable global merging).
- The MFGEA algorithm should be treated as a sample method from the group of methods using reduction targeted at structure simplification and increasing of accuracy.

5.6 Simulation Results

The comments on the simulations can be summarized as follows:

- The aim of the simulations was to show the mechanisms affecting fuzzy systems interpretability. A special focus was put on, showing how the reduction algorithms described in Sects. 5.4 and 5.5 work.
- The simulation problems discussed are shown in Table 5.4. The simulations were limited to the classification problems for which the reduction effect is more pronounced than for approximation problems. Two sample cases were considered.
- In the learning phase the error backpropagation algorithm described in Sect. 5.2 with $\eta = 0.25$ was used. The learning process took 1000 steps (generations). In the reduction phase it was assumed that $w_{\text{red}} = 1$ and $sim_{\text{mer}} = 0$. Moreover, it was assumed that the accuracy of the system after the reduction cannot be deteriorated in comparison to that from before the reduction.
- In the simulations the stratified cross-validation procedure described in Sect. 2.3.2 was used. The evaluation of the fuzzy systems accuracy was performed using the percentage classification error expressed by formula (2.20).
- In the simulations the system implementing the logical-type inference [47] of the form (4.28) with Dombi triangular norms and the Gaussian membership functions of the form (4.29) was used. The computations were performed for different combinations of the number of rules ($N = 1, \ldots, 4$) and discretization points ($R = 2, \ldots, 4$).
- The results of the simulations are presented in Tables 5.5 and 5.6 and in Figs. 5.12 and 5.13. They involve different combinations of the tested algorithms. The results are averaged in the context of the learning and testing sequences.
- The reduction level of the fuzzy system shown in Figs. 5.12a and 5.13a means the percentage reduction level of all the system (4.28) parameters mentioned in Sect. 4.3. Their number results from the number of fuzzy rules, input attributes, fuzzy sets and discretization points.
- The number of the used input attributes shown in Figs. 5.12b and 5.13b means the percentage usage level of input attributes in the system (4.28) after the reduction.

Table 5.4 The test problems used for the simulations

Item no.	Problem name	Number of input attributes	Number of output attributes	Number of sets	problem type	Problem label
1.	Glass identification problem [33]	9	2	214	Classification	GI
2.	Wisconsin breast cancer [33]	9	2	683	Classification	WBC

Table 5.5 A set of average, percentage classification errors of a not-reduced system and a reduced system of the form (4.28) for GI problem for a different number of fuzzy rules N and discretization points R

Item no.	Algorithms	$N =$	$R = 2$	$R = 3$	$R = 4$
1.	BP (not-reduced system)	1	5.88	5.86	5.68
2.	BP+CEA+CMA	1	5.81	5.56	5.38
3.	BP+MFLEA+CMA	1	5.10	5.10	5.38
4.	BP+MFGEA+CMA	1	4.57	4.82	4.92
5.	BP (not-reduced-system)	2	4.85	4.24	2.78
6.	BP+CEA+CMA	2	4.80	4.22	2.75
7.	BP+MFLEA+CMA	2	4.32	3.96	2.50
8.	BP+MFGEA+CMA	2	4.07	3.48	1.99
9.	BP (not-reduced-system)	3	3.56	2.52	2.17
10.	BP+CEA+CMA	3	3.54	2.50	2.15
11.	BP+MFLEA+CMA	3	3.28	2.27	2.15
12.	BP+MFGEA+CMA	3	2.80	2.02	1.89
13.	BP (not-reduced-system)	4	3.05	2.09	1.24
14.	BP+CEA+CMA	4	3.03	2.07	1.24
15.	BP+MFLEA+CMA	4	3.01	2.05	0.98
16.	BP+MFGEA+CMA	4	2.78	1.82	0.98

Table 5.6 A list of average, percentage classification errors of a not-reduced system and a reduced systems of the form (4.28) for the WBC problem for a different number of fuzzy rules N and discretization points R

Item no.	Algorithms	$N =$	$R = 2$	$R = 3$	$R = 4$
1.	BP (system not-reduced)	1	3.30	2.45	2.25
2.	BP+CEA+CMA	1	3.29	2.45	2.25
3.	BP+MFLEA+CMA	1	3.20	2.28	2.14
4.	BP+MFLEA+CMA	1	3.13	2.27	2.16
5.	BP (system not-reduced)	2	1.90	1.83	1.82
6.	BP+CEA+CMA	2	1.89	1.84	1.81
7.	BP+MFLEA+CMA	2	1.71	1.73	1.72
8.	BP+MFGEA+CMA	2	1.74	1.81	1.71
9.	BP (system not-reduced)	3	1.91	1.73	1.54
10.	BP+CEA+CMA	3	1.91	1.72	1.54
11.	BP+MFLEA+CMA	3	1.82	1.63	1.53
12.	BP+MFGEA+CMA	3	1.80	1.63	1.51
13.	BP (system not-reduced)	4	1.63	1.46	1.34
14.	BP+CEA+CMA	4	1.63	1.47	1.34
15.	BP+MFLEA+CMA	4	1.62	1.45	1.34
16.	BP+MFGEA+CMA	4	1.61	1.44	1.26

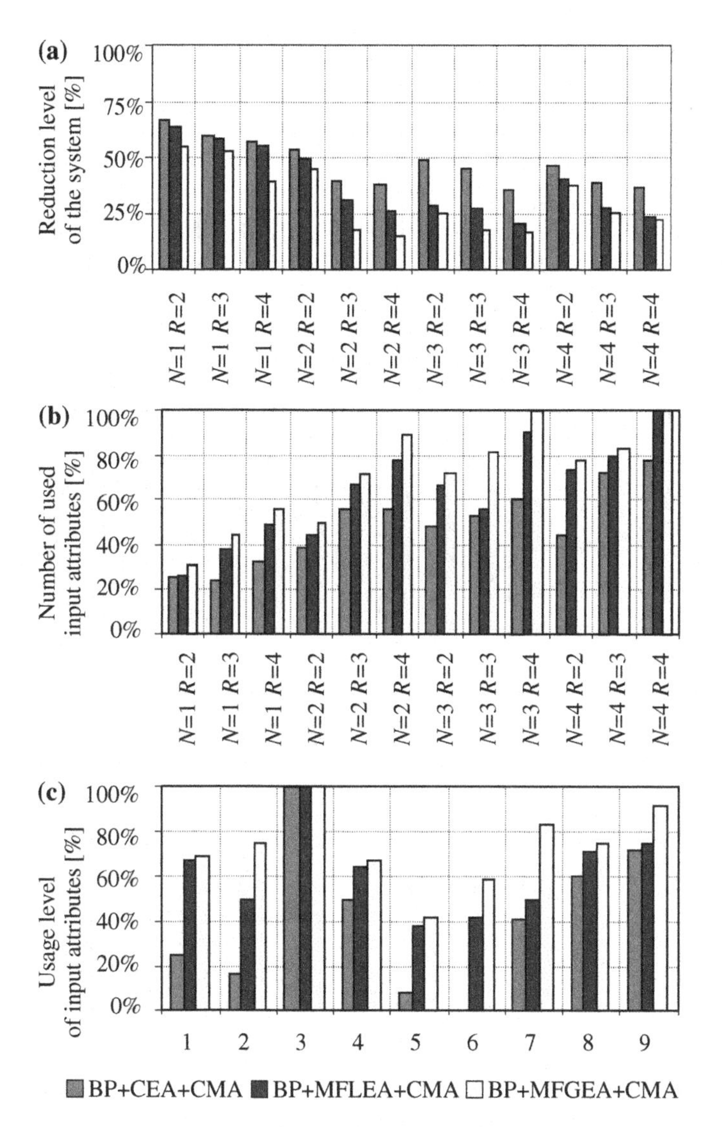

Fig. 5.12 Impact of reduction and merging on the structure of the fuzzy system of the form (4.28) for the GI problem: **a** the reduction level of the system, **b** the usage level of input attributes after the reduction, **c** the frequency of using input attributes in the group of the tested systems after the reduction

For example, the value of 60% means that the system (4.28) works without deterioration of the quality on the basis of 60% of the initial number of input attributes (i.e. before the reduction).

The conclusions on the simulations can be summarized as follows:

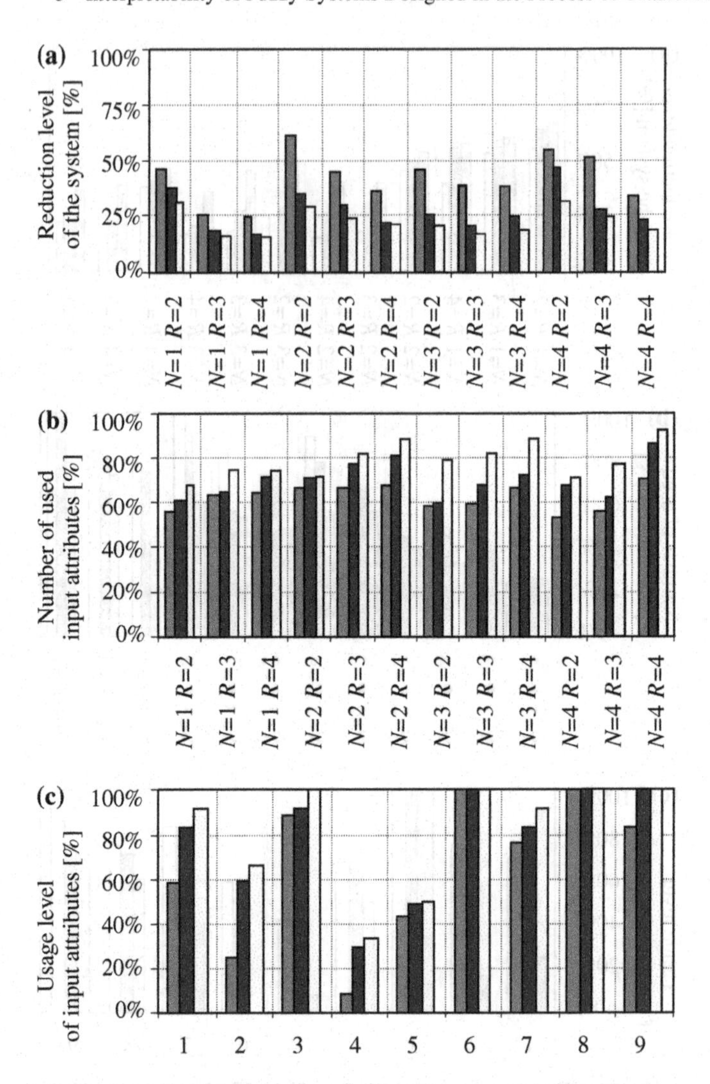

Fig. 5.13 Impact of reduction and merging on the structure of the fuzzy system of the form (4.28) for the WBC problem: **a** the reduction level of the system, **b** the usage level of input attributes after reduction, **c** the frequency of using input attributes in the group of the tested systems after the reduction

- Each presented reduction and merging algorithm works as expected, i.e. it results in a simplification of the fuzzy system (4.28) structure and does not deteriorate its accuracy (Tables 5.5 and 5.6).
- The CEA and CMA algorithms cause the most intense reduction of the fuzzy system (Figs. 5.12a and 5.13a). In some cases the use of these algorithms also

resulted in a slight improvement in the accuracy although that was not actually directly intended.

- The MFLEA and CMA algorithms result in an improvement in the accuracy of the system (Tables 5.5 and 5.6). The resulting reduction level of the system, however, is weaker (as expected).
- The MFGEA and CMA algorithms also cause an improvement in the accuracy of the system (Tables 5.5 and 5.6), like the MFLEA and CMA algorithms. An improvement in the accuracy is clearer than in the case of the MFLEA and CEA algorithms.
- The reduced systems can operate without compromising the accuracy even on the basis of a considerably limited subset of input attributes initial set (Figs. 5.12b and 5.13b). The number of the input attributes required for the operation depends on a simulation problem.
- Some input attributes were reduced in all or almost all discussed fuzzy systems (Figs. 5.12c and 5.13c).

5.7 Summary

In this chapter we discussed different ways of influencing interpretability of fuzzy systems designed in the process of gradient learning. In the introduction we describe the role of the initialization procedure (Sect. 5.1) and the basic gradient algorithm, i.e. the method of error backpropagation (Sect. 5.2). We indicate its limited possibilities for the selection of the fuzzy system structure (number of rules, number of input attributes, number of sets and number of discretization points). Therefore, the rest of this section presents the solutions supporting interpretable fuzzy systems designing, i.e. the reduction and merging mechanisms. They create good opportunities for the selection of the structure of an already learnt fuzzy system and they can operate without a negative impact on its accuracy. We present different approaches to controlling the reduction (Sects. 5.3 and 5.4) emphasizing that this mechanism can be freely adjusted to individual needs and can, for example, include interpretability criteria. We also emphasize flexibility of the assumptions about the selection mechanism of the components which are candidates for reduction. However, a common feature of the proposed algorithms is that they are based on immediate, practical verification of effects of an implemented reduction process. In the final part of this chapter (Sect. 5.5) the exemplary simulation results of the proposed reduction mechanisms are presented. The most important conclusion from the simulations is that the reduction mechanism can contribute to a significant simplification of the structure of a fuzzy system. This simplification has an impact on an increase of interpretability of a fuzzy system being reduced.

It is worth noting that the solutions from the field of reduction proposed in this chapter can be used for any computational intelligence systems whose structure has to be individually selected.

Some aspects of affecting interpretability of fuzzy systems designed by gradient learning were partially considered in our previous papers (see e.g. [9–11, 47]).

References

1. Aggarwal, C.C., Reddy, C.K.: Data Clustering: Algorithms and Applications. Chapman and Hall/CRC, Boca Raton (2013)
2. Albalate, A., Suendermann, D.: A combination approach to cluster validation based on statistical quantiles. In: Proceedings of the 2009 International Joint Conference on Bioinformatics, Systems Biology and Intelligent Computing (IJCBS'09), pp. 549–555 (2009)
3. Arabas, J., Kozdrowski, S.: Population initialization in the context of a biased, problem-specific mutation. In: Proceedings of the 1998 IEEE World Congress on Computational Intelligence, The 1998 IEEE International Conference on Evolutionary Computation, pp. 769–774 (1998)
4. Baccour, L., Alimi, A.M., John, R.I.: Some notes on fuzzy similarity measures and application to classification of shapes, recognition of arabic sentences and mosaic. Int. J. Comput. Sci. **41**, 81–90 (2014)
5. Beg, I., Ashraf, S.: Similarity measures for fuzzy sets. Appl. Comput. Math. **8**, 192–202 (2009)
6. Bezdek, J.C., Ehrlich, R., Full, W.: FCM: the fuzzy c-means clustering algorithm. Comput. Geosci. **10**, 191–203 (1984)
7. Casillas, J., Cordon, O., Herrera, F., Magdalena, L. (eds.): Interpretability Issues in Fuzzy Modeling. Springer, Heidelberg (2003)
8. Chua, T., Tan, W.: A new fuzzy rule-based initialization method for K-Nearest neighbor classifier. In: Proceedings of the 2009 IEEE International Conference on Fuzzy Systems (FUZZ-IEEE'09), pp. 415–420 (2009)
9. Cpałka, K.: A new method for design and reduction of neuro-fuzzy classification systems. IEEE Trans. Neural Network. **20**, 701–714 (2009)
10. Cpałka, K., Rutkowski, L.: Flexible neuro-fuzzy structures for pattern classification. WSEAS Trans. Comput. **4**, 679–688 (2005)
11. Cpałka, K., Rutkowski, L.: Flexible Takagi–Sugeno neuro-fuzzy structures for nonlinear approximation. WSEAS Trans. Syst. **4**, 1450–1458 (2005)
12. Das, A.K., Sil, J.: Cluster validation using splitting and merging technique. In: Proceedings of the 2007. International Conference on Computational Intelligence and Multimedia Applications, pp. 56–60 (2007)
13. Dasgupta, D., Hernandez, G., Romero, A., Garrett, D., Kaushal, A., Simien, J.: On the use of informed initialization and extreme solutions sub-population in multi-objective evolutionary algorithms. In: Proceedings of the IEEE Symposium on Computational Intelligence in Multi-criteria Decision-making, pp. 58–65 (2009)
14. Dubois, D., Prade, H., Esteva, F., Garcia, P., Godo, L.: A logical approach to interpolation based on similarity relations. Int. J. Approx. Reason. **17**, 1–36 (1997)
15. Everitt, B.S., Landau, S., Leese, M., Stahl, D.: Cluster Analysis. Wiley, Chichester (2011)
16. Gao, W.F., Liu, S.Y., Huang, L.L.: Particle swarm optimization with chaotic opposition-based population initialization and stochastic search technique. Commun. Nonlinear Sci. Numer. Simul. **17**, 4316–4327 (2012)
17. Graham, L., Oppacher, F.: Symmetric comparator pairs in the initialization of genetic algorithm populations for sorting networks. In: Proceedings of the 2006 IEEE International Conference0010890 on Evolutionary Computation, pp. 2845–2850 (2006)
18. Guillaume, S.: Designing fuzzy inference systems from data: an interpretability oriented review. IEEE Trans. Fuzzy Syst. **9**, 426–443 (2001)
19. Guillaume, S., Charnomordic, B.: Generating an interpretable family of fuzzy partitions from data. IEEE Trans. Fuzzy Syst. **12**, 324–335 (2004)

20. Gupta, K.R.: Effect of varying the size of initial parent pool in genetic algorithm. In: Proceedings of the 2014 International Conference on Contemporary Computing and Informatics (IC3I), pp. 785–788 (2014)
21. Halkidi, M., Batistakis, Y., Vazirgiannis, M.: On clustering validation techniques. J. Intell. Inf. Syst. **17**, 107–145 (2001)
22. Heaton, J.: Artificial Intelligence for Humans. Nature-Inspired Algorithms, vol. 2. CreateSpace Independent Publishing Platform, Dover (2014)
23. Icke, I., Rosenberg, A.: Multi-objective Genetic Programming for Visual Analytics. Lecture Notes in Computer Science, pp. 322–334. Springer, Heidelberg (2011)
24. Ishibuchi, H., Nakashima, T., Murata, T.: Performance evaluation of fuzzy classifier systems for multidimensional pattern classification problems. IEEE Trans. Syst. Man Cybern. Part B **29**, 601–618 (1999)
25. Jin, Y.: Fuzzy modeling of high dimensional systems: complexity reduction and interpretability improvement. IEEE Trans. Fuzzy Syst. **8**, 212–221 (2000)
26. Kazimipour, B., Li, X., Qin, A.K.: A review of population initialization techniques for evolutionary algorithms. In: Proceedings of the 2014 IEEE Congress on Evolutionary Computation (CEC), pp. 2585–2592 (2014)
27. Kazimipour, B., Li, X., Qin, A.K.: Effects of population initialization on differential evolution for large scale optimization. In: Proceedings of the 2014 IEEE Congress on Evolutionary Computation (CEC), pp. 2404–2411 (2014)
28. Kenesei, T., Abonyi, J.: Interpretable support vector machines in regression and classification - application in process engineering. Hung. J. Ind. Chem. **35**, 101–108 (2007)
29. Khan, S.S., Ahmad, A.: Cluster center initialization algorithm for K-modes clustering. Expert Syst. Appl. **40**, 7444–7456 (2013)
30. Krishnamoorthy, R., Kumar, S.S.: A new inter cluster validation method for unsupervised clustering techniques. In: Proceedings of the 2013 International Conference on Communication and Computer Vision (ICCCV), pp. 1–5 (2013)
31. Krishnamoorthy, R., Kumar, S.S.: Optimized cluster validation technique for unsupervised clustering techniques. In: Proceedings of the 2014 International Conference on Information Communication and Embedded Systems (ICICES), pp. 1–6 (2014)
32. Lee-Kwang, H., Song, T.S., Lee, K.M.: Similarity measure between fuzzy sets and between elements. Fuzzy Sets Syst. **62**, 291–293 (1994)
33. Lichman, M.: UCI Machine Learning Repository. Irvine, CA: University of California, School of Information and Computer Science (2013). http://archive.ics.uci.edu/ml
34. Liu, F., Quek, C., Ng, G.S.: A novel generic hebbian ordering based fuzzy rule base reduction approach to Mamdani neuro fuzzy system. Neural Comput. **19**, 1656–1680 (2007)
35. Mishra, S., Saha, S., Mondal, S.: Cluster validation techniques for Bibliographic databases. In: Proceedings of the 2014 IEEE Students' Technology Symposium (TechSym), pp. 93–98 (2014)
36. Mishra, S., Saha, S., Mondal, S.: On validation of clustering techniques for bibliographic databases. In: Proceedings of the 2014 22nd International Conference on Pattern Recognition (ICPR), pp. 3150–3155 (2014)
37. Modiri-Delshad, M., Rahim, N.A.: Fast initialization of population based methods for solving economic dispatch problems. In: Proceedings of the 3rd IET International Conference on Clean Energy and Technology (CEAT), pp. 1–5 (2014)
38. Nauck, D., Kruse, R.: Designing neuro-fuzzy systems through back-propagation. In: Pedrycz, W. (ed.) Fuzzy Modeling: Paradigms and Practice, pp. 203–228. Kluwer Academic Publishers, Norwell (1996)
39. Orito, Y., Hanada, Y., Shibata, S., Yamamoto, H.: A new population initialization approach based on bordered hessian for portfolio optimization problems. In: Proceedings of the 2013 IEEE International Conference on Systems, Man, and Cybernetics, pp. 1341–1346 (2013)
40. Paul, P.V., Dhavachelvan, P., Baskaran, R.: A novel population initialization technique for genetic algorithm, circuits. In: Proceedings of the 2013 International Conference on Power and Computing Technologies (ICCPCT), pp. 1235–1238 (2013)

41. Poikolainen, I., Neri, F., Caraffini, F.: Cluster-based population initialization for differential evolution frameworks. Inf. Sci. **297**, 216–235 (2015)
42. Rahnamayan, S., Tizhoosh, H.R., Salama, M.M.A.: A novel population initialization method for accelerating evolutionary algorithms. Comput. Math. Appl. **53**, 1605–1614 (2007)
43. Riid, A., Rustern, E.: Interpretability improvement of fuzzy systems: reducing the number of unique singletons in zeroth order Takagi–Sugeno systems. In: Proceedings of the IEEE International Conference on Fuzzy Systems, pp. 1–6 (2010)
44. Roubos, H., Setnes, M.: Compact and transparent fuzzy models and classifiers through iterative complexity reduction. IEEE Trans. Fuzzy Syst. **9**, 516–524 (2001)
45. Rutkowski, L.: Flexible Neuro-Fuzzy Systems: Structures, Learning and Performance Evaluation. Springer, Heidelberg (2004)
46. Rutkowski, L.: Computational Intelligence. Springer, Heidelberg (2010)
47. Rutkowski, L., Cpałka, K.: Designing and learning of adjustable quasi-triangular norms with applications to neuro-fuzzy systems. IEEE Trans. Fuzzy Syst. **13**, 140–151 (2005)
48. Sekaj, I., Perkacz, J.: Some aspects of parallel genetic algorithms with population re-initialization. In: Proceedings of the 2007 IEEE Congress on Evolutionary Computation, pp. 1333–1338 (2007)
49. Shukla, P.K., Tripathi, S.P.: Handling high dimensionality and interpretability-accuracy trade-off issues in evolutionary multiobjective fuzzy classifiers. Int. J. Sci. Eng. Res. **5**, 665–671 (2014)
50. Stetco, A., Zeng, X.J., Keane, J.: Fuzzy C-means++: fuzzy C-means with effective seeding initialization. Expert Syst. Appl. **42**, 7541–7548 (2015)
51. Tadeusiewicz, R.: Neural networks (in Polish). Exit (1993)
52. Wang, H., Wu, Z., Wang, J., Dong, X., Yu, S., Chen, C.: A new population initialization method based on space transformation search. In: Proceedings of the 2009 Fifth International Conference on Natural Computation, pp. 332–336 (2009)
53. Werbos, P.: Beyond regression: New tools for prediction and analysis in the behavioral sciences (Ph.D. thesis). Harvard University (1974)
54. Werbos, P.: Applications of advances in nonlinear sensitivity analysis. In: System Modeling and Optimization. Springer (1982)
55. Xu, R., Wunsc, D.: Clustering. Wiley-IEEE Press, New York (2008)
56. Yong, Z., Li-min, J., Wei-l, H.: Construct interpretable fuzzy classification system based on fuzzy clustering initialization. Int. J. Inf. Technol. **11**, 91–107 (2005)
57. Zhang, Y., Yang, R., Zuo, J., Jing, X.: Enhancing MOEA/D with uniform population initialization, weight vector design and adjustment using uniform design. J. Syst. Eng. Electron. **26**, 1010–1022 (2015)
58. Zwick, R., Carlstein, E., Budescu, D.V.: Measures of similarity among fuzzy concepts: a comparative analysis. Int. J. Approx. Reason. **1**, 221–242 (1987)

Chapter 6
Interpretability of Fuzzy Systems Designed in the Process of Evolutionary Learning

Chapter 5 presents possible ways of affecting interpretability of fuzzy systems designed by using gradient learning. Gradient learning is an alternative to evolutionary learning, which is implemented using population-based algorithms. A variety of subjects associated with those algorithms are often discussed in the literature [1, 2, 5, 7–11, 13–15, 19, 21, 23, 24, 26, 28, 29, 32, 39, 40, 46–48, 51, 57–59, 63, 66]. New methods of initialization and management of population, new exploration and exploitation operators of consideration space, new ways of encoding of particular problems, etc. are being created all the time.

A great interest in evolutionary learning results from the advantages offered by population-based algorithms, which are searching procedures mainly used to solve optimization problems. Their operation is often inspired by various forms of evolution (e.g. natural, social, etc.). Evolutionary learning bases on mechanisms of natural selection, inheritance, competition, etc. Moreover, it uses the concept of survival of the fittest individuals from the theory of evolution. Population-based algorithms differ from traditional optimization methods in a number of ways, which also include the following:

- They conduct a search starting not from a single point, but from their population. This approach makes it easier to find a global optimum during a learning process and increases the resistance of the algorithm to an incorrectly performed initialization.
- They do not directly process the parameters of a task but their encoded form. This allows them to operate independently of the problem under consideration and allows them to be easily adjusted to solve a variety of problems.
- They only use the objective function, not its derivatives. This feature also contributes to the independence of the operation of population-based algorithms from the problem under consideration. Moreover, it increases flexibility when formulating a problem and evaluating its solutions.

K. Cpałka, *Design of Interpretable Fuzzy Systems*, Studies in Computational Intelligence 684, DOI 10.1007/978-3-319-52881-6_6

- They use probabilistic, not deterministic selection rules. This feature provides high flexibility when selecting algorithm architecture. This facilitates implementing mechanisms inspired by the surrounding reality (in a simplified form) when designing algorithms. What also results from this feature is an ability to make a compromise between the exploration and exploitation of the search space.

All those features give population-based algorithms an advantage over other optimization techniques such as analytical methods, enumerative methods, random methods, etc. [55]. However, in the context of designing interpretable fuzzy systems a specific type of problem is considered. One of its characteristic features includes the fact that the solution being sought represents a certain structure (e.g. the structure of a fuzzy system) and that this structure has a set of parameters (e.g. the parameters of a fuzzy system) dependent on that structure. The form of the structure depends on, among others, the complexity of a given problem and in the traditional approach it is usually chosen by the trial and error method. Therefore, population-based algorithms are now expected to facilitate automatic selection of the structure of a given fuzzy system (i.e. a "frame" of the solution) and its parameters (i.e. parameters of the "frame") while taking into account a specified set of expectations (e.g. requirements formulated in the form of interpretability criteria).

However, when using population-based algorithms in designing interpretable fuzzy systems we should remember about certain difficulty which results from different types of encoding (binary or real) and their dedicated evolutionary operators used by different algorithms. Meanwhile, the parameters of a fuzzy system are usually of the real type and the most convenient way for its structure encoding is using binary values. This is due to the need to include information on whether in a given structure there are specified connections, specified rules are active, signals associated with specified input attributes should be processed, etc. That is why it is difficult to directly use the existing variety of population-based algorithms in designing interpretable fuzzy systems. Hence, in this chapter we consider hybrid-type population-based algorithms, which are able to search for the structure of a fuzzy system and its parameters simultaneously. They are based on a hybrid method of encoding and hybrid evolutionary operators.

This chapter presents the issues of population initialization (Sect. 6.1), evolutionary learning of fuzzy systems (Sect. 6.2) and a sample way of accounting for interpretability criteria of fuzzy systems (Sect. 6.3) in an evolutionary learning process. The final part of this chapter contains exemplary simulation results (Sect. 6.4), its summary (Sect. 6.5) and bibliography.

6.1 Initializing a Population in Evolutionary Learning

Proper initialization of a population in the evolutionary learning can facilitate finding a global optimum and reduce computational effort connected with implementation of a searching procedure. When a population has a finite number of individuals, the initialization becomes more significant when it comes to solving problems of a number of variables. Such problem is e.g. the problem of selection of the structure and parameters of fuzzy system, which is considered in this section.

The methods of initialization of individuals in a population which prove to have worked well are equivalents of the methods discussed in Sect. 5.1. They mainly involve uniform initialization taking into account the area of the problem under consideration, various techniques of covering promising areas (i.e. containing the global optimum or local optima that meet specified requirements), as well as an appropriate use of the pseudo-random number generator. Some other initialization methods adapted to specific population-based algorithms have also been created (e.g. the opposition-based population initialization [54]).

The literature on the subject offers different criteria for division of initialization methods. For example, in paper [36] the following criteria are presented:

- Randomness. This category comprises stochastic and deterministic techniques. In stochastic techniques the result of an initialization process depends on the so-called stochastic initializers. They are the source of randomness. This group includes techniques based on the use of pseudo-random number generators and chaotic number generators. Whereas, in deterministic techniques the initialization result depends on the so-called deterministic initializers. They are designed to ensure uniform distribution of values in the entire search space. This group includes techniques based on the use of pseudo-random sequences and so-called uniform experimental design. They are specific algorithms providing a uniform distribution of random values in the whole search space.
- Compositionality. One-stage and multi-stage techniques belong to this category. In one-stage techniques the initialization process cannot be divided into separable stages, whereas the group of multi-stage techniques contains hybrid and multi-step techniques.
- Generality. This category comprises generic and application specific techniques. In generic techniques we assume that we do not have knowledge about a given problem which is treated as a black box. Therefore, it is not possible to narrow the search space of the solution by careful omission of the areas in which there is no global optimum. The other type of these techniques - application specific techniques - use knowledge about the problem being considered in order to target initialization of a given population at covering particularly promising areas.

Moreover, the issue of initialization of a population in the evolutionary learning may include additional operations which are related to:

- Automatic selection of the population size and flexible adjustment of the size to the process of evolution.
- Automatic re-initialization (exchange) of a certain part of a population in the evolution process.
- Division of a population between different initialization techniques and aggregation of individuals generated in such process.

More information about initialization of fuzzy systems can be found in the literature (see e.g. [4, 20, 27, 30, 31, 33, 36–38, 49, 50, 52–54, 56, 62, 64]).

6.2 Evolutionary Learning of Fuzzy Systems

General comments on population-based algorithms in relation to their application in the evolutionary learning of fuzzy systems can be summarized as follows:

- They are well suited for selection of continuous parameters of a fuzzy system which have no limitations. Thus, they can be used when selecting fuzzy sets parameters, shape parameters of parameterized triangular norms, discretization points, etc.
- They can be used when learning continuous parameters of a fuzzy system which have limitations. Such parameters are weights of importance, shape parameters of some parameterized triangular norms (Tables 4.1 and 4.2), the inference-type parameter, etc. In such case we can use one of the mechanisms preventing the values of these parameters from falling outside the allowed range. This mechanism can be based on the use of range functions and it is described in Sect. 5.2. It can also base on the population recovery procedure, which is a typical mechanism used in population-based algorithms. It involves control of the range of the parameters encoded in individuals of a given population. If, after using evolutionary operators, the value of such parameters exceeds the allowed range, then their value is set as an appropriately determined value from the expected range. An interesting way of controlling such range is the use of the approach based on sets of values presented in Sect. 4.5.
- They make it possible to process parameters which take their values from sets. This feature is particularly important because it facilitates designing of a system, makes selection of the system structure and problem description flexible and ensures operational accuracy of a system. This property also allows us to automatically choose the number of fuzzy system rules, the number of fuzzy sets, the operator type (aggregation, inference and defuzzification), the type of membership functions, to select a subset of key input attributes of the system, etc. It is an important feature in the context of interpretable fuzzy systems design, because it can reduce redundancy in the system structure as early as during the learning stage. Using this approach requires an appropriate encoding of individuals (e.g. fuzzy systems) in a given population as well as suitable processing.
- They facilitate accounting for different interpretability assumptions (which are presented in detail in Sects. 3.4 and 4.4) and a proper compromise between the accuracy of a system and its interpretability. This property is particularly important from the point of view of interpretable fuzzy systems design.

Further on this chapter describes an example of two different hybrid population-based algorithms (Sects. 6.2.1 and 6.2.2). They represent two groups of population-based algorithms discussed in Sect. 2.2: one-population and multi-population algorithms. Hybrid population-based algorithms have been created by combining two types of algorithms:

- The algorithm using a binary encoding. Its use allows us to properly encode and process the structure of a fuzzy system. In this chapter the algorithm is assumed to be a genetic algorithm [34].
- The algorithm using a real encoding. Its use allows us to properly encode and process th parameters of a fuzzy system structure. In Sect. 6.2.1 the algorithm is assumed to be a firework algorithm [60, 65], whereas in Sect. 6.2.2 the algorithm is an imperialist algorithm [6].

In our previous papers we also discussed other combinations of base algorithms [16–18, 41–45]. The development of a hybrid algorithm requires appropriate encoding, ensuring synchronization of its components and determining an appropriate balance between exploration and exploitation of the consideration space.

6.2.1 Evolutionary Learning of Fuzzy Systems Using Single Population

This section shows how to perform the evolutionary learning of a fuzzy system with the use of a sample hybrid algorithm using a single population. The algorithm was created by combining the genetic algorithm [34] and the firework algorithm [60, 65]. From now on it will be referred to as the GFA. The genetic algorithm is meant to select the structure of a fuzzy system while the firework algorithm is meant to select its parameters. The idea of the genetic algorithm is inspired by the biological evolution of species while the idea of the firework algorithm by behavior of exploding fireworks. In the artificial firework algorithm "fireworks" represent potential solutions. When they "shoot up", they randomly cover the search space of considerations. Then, each of them generates "sparks", interpreted as solutions grouped around corresponding "fireworks". After the "fireworks" and the corresponding "sparks" are generated, their evaluation is performed. Then, the best of them are chosen to be taken to the next step of the algorithm. Repeating such action a specified number of times increases real chances of obtaining solutions close to the optimum (i.e. adopted) evaluation criteria (Fig. 6.1).

6.2.1.1 Encoding of Potential Solutions in a Population

Encoding of a population of potential solutions used in the algorithm refers to the Pittsburgh approach [35]. A single individual of the population $\mathbf{X}_{ch}$ thus encodes a single fuzzy system of the form (4.28) (or other form):

$$\mathbf{X}_{ch} = \left\{\mathbf{X}_{ch}^{\mathrm{str}}, \mathbf{X}_{ch}^{\mathrm{par}}\right\}. \tag{6.1}$$

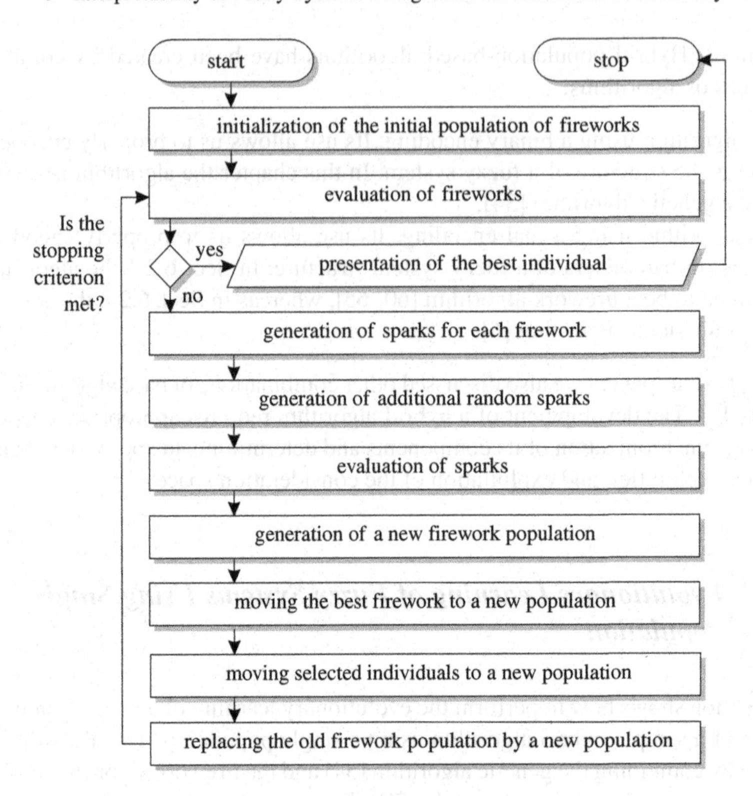

Fig. 6.1 Block scheme of the hybrid genetic-firework algorithm

Part $\mathbf{X}_{ch}^{\mathrm{str}}$ of the individual $\mathbf{X}_{ch}$ encodes the structure of a fuzzy system (4.28) using the binary approach and has the following form:

$$\mathbf{X}_{ch}^{\mathrm{str}} = \left\{ \begin{array}{c} x_1, \ldots, x_n, \\ A_1^1, \ldots, A_n^1, \ldots, A_1^{Nmax}, \ldots, A_n^{Nmax}, \\ B_1^1, \ldots, B_m^1, \ldots, B_1^{Nmax}, \ldots, B_m^{Nmax}, \\ rule_1, \ldots, rule_{Nmax}, \\ \bar{y}_{1,1}^{\mathrm{def}}, \ldots, \bar{y}_{1,Rmax}^{\mathrm{def}}, \ldots, \bar{y}_{m,1}^{\mathrm{def}}, \ldots, \bar{y}_{m,Rmax}^{\mathrm{def}} \end{array} \right\} = \left\{ X_{ch,1}^{\mathrm{str}}, \ldots, X_{ch,L^{\mathrm{str}}}^{\mathrm{str}} \right\}, \quad (6.2)$$

where $ch = 1, \ldots, Npop$ is the index of an individual in a population, $Npop$ is the number of individuals in a population, $Nmax$ is the maximum (allowed) number of rules in the system (4.28) (selected individually for a given problem), $Rmax$ is the maximum (allowed) number of discretization points in the system (4.28) (also selected individually for each considered problem), and L^{str} is the number of the individual components $\mathbf{X}_{ch}^{\mathrm{str}}$ (from now on referred to as genes, like in the genetic algorithm), which is determined as follows:

$$L^{\mathrm{str}} = Nmax \cdot (n + m + 1) + n + Rmax \cdot m. \quad (6.3)$$

In the encoding procedure of $\mathbf{X}_{ch}^{\mathrm{str}}$ it is assumed that each individual of the population encodes the maximum number of rules $Nmax$ indicated by the user and the number of discretization points $Rmax$ independently for each system output. When operating the algorithm searches the real number of the system (4.28) rules in the range $N \in [1, Nmax]$ and the real number of discretization points in the range $R_j \in [1, Rmax]$ $(j = 1, \ldots, m)$. Thus, this is a different approach from the one considered in Sect. 5.2, in which the user (mostly using the trial-and-error method) had to clearly indicate N and R_j (as redundant components of the fuzzy system can be removed by reduction on completion of the learning process).

The principle adopted in the encoding procedure $\mathbf{X}_{ch}^{\mathrm{str}}$ is such that the gene with value 0 of the individual $\mathbf{X}_{ch}^{\mathrm{str}}$ excludes the associated element from the target system structure (4.28) and vice versa. This element can have different forms including a rule ($rule_k$, $k = 1, \ldots, Nmax$), an input fuzzy set (A_i^k, $i = 1, \ldots, n$, $k = 1, \ldots, Nmax$), an output fuzzy set ($B_j^k, j = 1, \ldots, m, k = 1, \ldots, Nmax$), an input attribute ($\bar{x}_i$, $i = 1, \ldots, n$), or a discretization point ($\bar{y}_{j,r}^{\mathrm{def}}$, $r = 1, \ldots, Rmax$).

Part $\mathbf{X}_{ch}^{\mathrm{par}}$ of the individual $\mathbf{X}_{ch}$ encodes the real parameters of the fuzzy system of the form (4.28). Form $\mathbf{X}_{ch}^{\mathrm{par}}$ is dependent on detailed principles of the system. The remaining part of this chapter contains considerations on a fuzzy system of the form (4.28) with Gaussian membership functions (4.29) and aggregation and inference operators built on the basis of parameterized Dombi-type triangular norms of the form (4.1). Then, $\mathbf{X}_{ch}^{\mathrm{par}}$ has the following form:

$$\mathbf{X}_{ch}^{\mathrm{par}} = \left\{ \begin{array}{c} \bar{x}_{1,1}^{A}, \sigma_{1,1}^{A}, \ldots, \bar{x}_{n,1}^{A}, \sigma_{n,1}^{A}, \ldots \\ \bar{x}_{1,Nmax}^{A}, \sigma_{1,Nmax}^{A}, \ldots, \bar{x}_{n,Nmax}^{A}, \sigma_{n,Nmax}^{A}, \\ \bar{y}_{1,1}^{B}, \sigma_{1,1}^{B}, \ldots, \bar{y}_{m,1}^{B}, \sigma_{m,1}^{B}, \ldots \\ \bar{y}_{1,Nmax}^{B}, \sigma_{1,Nmax}^{B}, \ldots, \bar{y}_{m,Nmax}^{B}, \sigma_{m,Nmax}^{B}, \\ w_{1,1}^{A}, \ldots, w_{1,n}^{A}, \ldots, w_{Nmax,1}^{A}, \ldots, w_{Nmax,n}^{A}, \\ w_{1,1}^{B}, \ldots, w_{m,1}^{B}, \ldots, w_{1,Nmax}^{B}, \ldots, w_{m,Nmax}^{B}, \\ w_{1}^{\mathrm{rule}}, \ldots, w_{Nmax}^{\mathrm{rule}}, \\ p^{\tau}, p^{\mathrm{imp}}, p^{\mathrm{agr}}, \\ \bar{y}_{1,1}^{\mathrm{def}}, \ldots, \bar{y}_{1,Rmax}^{\mathrm{def}}, \ldots, \bar{y}_{m,1}^{\mathrm{def}}, \ldots, \bar{y}_{m,Rmax}^{\mathrm{def}} \end{array} \right\} = \left\{X_{ch,1}^{\mathrm{par}}, \ldots, X_{ch,L^{\mathrm{par}}}^{\mathrm{par}}\right\}, \tag{6.4}$$

where $\left\{\bar{x}_{i,k}^{A}, \sigma_{i,k}^{A}\right\}$ are the parameters of the Gaussian membership function (4.29) of input fuzzy sets $A_1^k, \ldots, A_n^k$, $\left\{\bar{y}_{j,k}^{B}, \sigma_{j,k}^{B}\right\}$ are the parameters of the Gaussian membership function (4.29) of output fuzzy sets $B_1^k, \ldots, B_m^k$, $\left\{p^{\tau}, p^{\mathrm{imp}}, p^{\mathrm{agr}}\right\}$ are the shape parameters of parameterized Dombi-type triangular norms and L^{par} is the number of components of individual $\mathbf{X}_{ch}^{\mathrm{par}}$, which is determined as follows:

$$L^{\mathrm{par}} = Nmax \cdot (3 \cdot n + 3 \cdot m + 1) + Rmax \cdot m + 3. \tag{6.5}$$

Interpretation of the other parameters encoded in $\mathbf{X}_{ch}^{\mathrm{par}}$ is presented in the context of the description of the fuzzy system of the form (4.28).

In the encoding procedure of $\mathbf{X}_{ch}^{\mathrm{par}}$ it is assumed that only genes $\mathbf{X}_{ch}^{\mathrm{par}}$, whose counterparts in $\mathbf{X}_{ch}^{\mathrm{str}}$ are equal to 1, are considered in the construction of the fuzzy system (4.28). Moreover, when analyzing $\mathbf{X}_{ch}^{\mathrm{str}}$ of the form (6.2), it is easy to define the function used for determination of the real number of input attributes in the individual $\mathbf{X}_{ch}$:

$$\mathrm{noi}\,(ch) = \sum_{i=1}^{n} \mathbf{X}_{ch}^{\mathrm{str}}\,\{x_i\}, \tag{6.6}$$

where $\mathbf{X}_{ch}^{\mathrm{str}}\,\{x_i\}$ is the parameter (gene) of the individual $\mathbf{X}_{ch}^{\mathrm{str}}$ associated with the input x_i. The adoption of this notation greatly facilitated, among others, the notation of the interpretability criteria considered in Sect. 6.3. Similar to noi (ch), other functions used to determine the real number of fuzzy rules, fuzzy sets, discretization points, etc., can be defined. They are presented in Table 6.1 and used in Sect. 6.3.

The way of encoding potential solutions in the population described in this section can be easily adapted to fuzzy systems other than the system of the form (4.28).

6.2.1.2 Evaluation of the Population

As already mentioned, each individual $\mathbf{X}_{ch}$ encodes in the part $\mathbf{X}_{ch}^{\mathrm{par}}$ the parameters (formula (6.4)) and in the part $\mathbf{X}_{ch}^{\mathrm{str}}$ the structure (formula (6.2)) of a single fuzzy system (e.g. of the form (4.28)). It has been assumed that the purpose of the algorithm is to minimize the value of the evaluation function specified for the individual $\mathbf{X}_{ch}$ in the following way:

$$\mathrm{ff}\,(\mathbf{X}_{ch}) = T_a^* \left\{ \begin{array}{c} \mathrm{ffacc}\,(\mathbf{X}_{ch})\,,\, \mathrm{ffint}\,(\mathbf{X}_{ch})\,; \\ w_{\mathrm{ffacc}},\, w_{\mathrm{ffint}} \end{array} \right\}, \tag{6.7}$$

where the component ffacc $(\mathbf{X}_{ch})$ specifies the accuracy of the system (4.28), the component ffint $(\mathbf{X}_{ch})$ specifies the interpretability of a system (4.28) (according to the adopted interpretability criteria), $T_a^*\,\{\cdot\}$ is the algebraic t-norm with weights of the form (4.8), $w_{\mathrm{ffacc}} \in [0, 1]$ represents the weight of the component ffacc $(\mathbf{X}_{ch})$ and $w_{\mathrm{ffint}} \in [0, 1]$ represents the weight of the component ffint $(\mathbf{X}_{ch})$. The values of weights w_{ffacc} i w_{ffint} result from expectations of the user regarding the ratio between the accuracy of the system (4.28) and its interpretability.

The way in which the component ffacc $(\mathbf{X}_{ch})$ is determined depends on the type of the problem being considered (Sect. 2.3). In the case of approximation issues we can use the formula (2.19), and in the case of classification issues we can use the formula (2.20), whereas the component ffint $(\mathbf{X}_{ch})$ can take into account interpretability criteria proposed in Sect. 6.3. Their aggregation can be conducted as follows:

Table 6.1 A set of functions used to determine the number of components in the fuzzy system (4.28) encoded in the individual $\mathbf{X}_{ch}$ of the form (6.1)

Item no.	Function name	Function formula
1.	Function used to determine the number of input attributes	$\text{noi}(ch) = \sum_{i=1}^{n} \mathbf{X}_{ch}^{\text{str}}\{x_i\}$
2.	Function used to determine the number of fuzzy rules	$\text{nor}(ch) = \sum_{k=1}^{Nmax} \mathbf{X}_{ch}^{\text{str}}\{rule_k\}$
3.	Function used to determine the number of input fuzzy sets for the input i	$\text{noifsfi}(ch, i) = \mathbf{X}_{ch}^{\text{str}}\{x_i\} \cdot \sum_{k=1}^{Nmax} \mathbf{X}_{ch}^{\text{str}}\{rule_k\} \cdot \mathbf{X}_{ch}^{\text{str}}\{A_i^k\}$
4.	Function used to determine the number of input fuzzy sets for the rule k	$\text{noifsir}(ch, k) = \mathbf{X}_{ch}^{\text{str}}\{rule_k\} \cdot \sum_{i=1}^{n} \mathbf{X}_{ch}^{\text{str}}\{x_i\} \cdot \mathbf{X}_{ch}^{\text{str}}\{A_i^k\}$
5.	Function used to determine the number of all input fuzzy sets	$\text{noifs}(ch) = \sum_{k=1}^{Nmax} \text{noifsir}(ch, k) = \sum_{i=1}^{n} \text{noifsfi}(ch, i)$
6.	Function used to determine the number of output fuzzy sets for the output j	$\text{noofsfo}(ch, j) = \sum_{k=1}^{Nmax} \mathbf{X}_{ch}^{\text{str}}\{rule_k\} \cdot \mathbf{X}_{ch}^{\text{str}}\{B_j^k\}$
7.	Function used to determine the number of output fuzzy sets for the rule k	$\text{noofsir}(ch, k) = \mathbf{X}_{ch}^{\text{str}}\{rule_k\} \cdot \sum_{j=1}^{m} \mathbf{X}_{ch}^{\text{str}}\{B_j^k\}$
8.	Function used to determine the number of all output fuzzy sets	$\text{noofs}(ch) = \sum_{k=1}^{Nmax} \text{noofsir}(ch, k) = \sum_{j=1}^{m} \text{noofsfi}(ch, j)$
9.	Function used to determine the number of discretization points for the output j	$\text{nodfo}(ch, j) = \sum_{r=1}^{Rmax} \mathbf{X}_{ch}^{\text{str}}\{\bar{y}_{j,r}^{\text{def}}\}$
10.	Function used to determine the number of all discretization points	$\text{nod}(ch) = \sum_{j=1}^{m} \text{nodfo}(ch, j)$

$$\text{ffint}(\mathbf{X}_{ch}) = T_a^* \left\{ \begin{array}{c} \text{ffint}_A(\mathbf{X}_{ch}), \text{ffint}_B(\mathbf{X}_{ch}), \ldots; \\ w_{\text{ffintA}}, w_{\text{ffintB}}, \ldots \end{array} \right\}, \tag{6.8}$$

where $\text{ffint}_A(\cdot)$, $\text{ffint}_B(\cdot)$,... are the functions representing the adopted interpretability criteria (e.g. those defined in Sect. 6.3), $w_{\text{ffintA}} \in [0, 1]$, $w_{\text{ffintB}} \in [0, 1]$,... are the weights of importance of the function $\text{ffint}_A(\cdot)$, $\text{ffint}_B(\cdot)$,...and $T_a^*\{\cdot\}$ is the algebraic t-norm with weights of the form (4.8). The values of the weights in equation (6.8) can be selected on the basis of the suggestions given in the paper [3].

In most applications, the objective in the design phase is to obtain a single fuzzy system. Such system is expected to have good accuracy and interpretability. However, if it was necessary to obtain a set of solutions with different accuracy-interpretability trade-off (in terms of Eq. (6.7)), then the possibilities offered by the methods based on the Pareto fronts [22] could be used instead of criteria aggregation.

6.2.1.3 Processing of the Population

The hybrid genetic-firework algorithm considered in this section implements actions typical for population-based algorithms [55]. Therefore, its operation can be divided into the following steps: initialization of the population, evaluation of the population, use of exploration and exploitation operators of the search space, test of the stopping criterion and creation of a new population. A block diagram of the algorithm is shown in Fig. 6.1. Its subsequent steps are described later in this section.

Step 1. Initiation of the Population

Initialization of the population refers to the individuals $\mathbf{X}_{ch}$, $ch = 1, \ldots, Npop$ ($Npop$ is the number of individuals in a population). These individuals are interpreted as fireworks, which actually are locations of their "explosion". In the basic initialization procedure of the population each gene $X^{\text{str}}_{ch,g}$ (affecting the form of the fuzzy system structure (4.28)) can be initially drawn from the set $\{0, 1\}$, and each gene $X^{\text{par}}_{ch,g}$ (determining the values of the structure parameters) can be initially drawn taking into account possible limitations on its value. In the initialization procedure we can also use other methods, e.g. the ones mentioned in Sect. 6.1.

Step 2. Evaluation of Population and Checking the Stopping Criterion

Evaluation of $Npop$ individuals from a population is implemented using the evaluation function ff ($\mathbf{X}_{ch}$) of the form (6.7). This procedure also selects the individual **XBest** which has the best value of the evaluation function (6.7). Moreover, its index $chBest$ is stored (**XBest** is $\mathbf{X}_{chBest}$).

The stopping criterion of the algorithm can involve performing a specified number of steps. This is the most typical condition, but others are also possible. They can involve e.g. achieving an adopted threshold value by using the evaluation function of the best individual in the population or calling the evaluation function a specified number of times. If the adopted stopping criterion is not satisfied, then the algorithm proceeds to the next step. If the stopping criterion is met, then the algorithm presents the information on the best individual in the population, i.e. individual **XBest**.

Step 3. Generation of Sparks

In this step fireworks generate sparks in order to exploit the search space. Only those individuals $\mathbf{X}^{\text{par}}_{ch}$ which encode parameters are modified in this step. The individuals $\mathbf{X}^{\text{str}}_{ch}$ encoding the structure are not modified - their modification is performed in the next step. If the individuals encoding the structure are modified with the same intensity as the ones encoding parameters, the algorithm has a small chance to find a satisfactory solution, i.e. an appropriate structure and parameters of a fuzzy system. The approach proposed in the considered algorithm indicates a set of structures encoded in individuals from a population and then looks for a solution in their surrounding. However, the mechanism for random sparks generation (described in the next step) allows us to search for new structures of the system (4.28). It is aimed at the exploration of the consideration space.

The operation of sparks generation discussed in this step involves determination of the number of sparks for each firework. This number is dependent on the value of the firework fitness function (also known as the adaptation function) (6.7). If the value of this firework function is smaller (in the minimization problem), the number of its sparks is greater. The initial number of sparks for the firework $\mathbf{X}_{ch}$ is determined as follows:

$$s_{ch} = Nspa \cdot \frac{\max\left\{\mathrm{ff}\left(\mathbf{X}_1\right), \ldots, \mathrm{ff}\left(\mathbf{X}_{Npop}\right)\right\} - \mathrm{ff}\left(\mathbf{X}_{ch}\right) + \xi}{\sum_{i=1}^{Npop}\left(\max\left\{\mathrm{ff}\left(\mathbf{X}_1\right), \ldots, \mathrm{ff}\left(\mathbf{X}_{Npop}\right)\right\} - \mathrm{ff}\left(\mathbf{X}_i\right)\right) + \xi}, \tag{6.9}$$

where $Nspa$ is the parameter of the algorithm which controls the number of created sparks and $\xi > 0$ is a constant small real number which prevents dividing by 0. The algorithm assumes that the target number of sparks for each firework (denoted as $\hat{s}_{ch}$) has to be within the range $\left[b \cdot Nspa, a \cdot Nspa\right]$, while the parameters a and b of the algorithm should meet the assumption $a < b < 1$. The value $a \cdot Nspa$ is the minimum number of generated sparks (the value a has to be relatively low), value $b \cdot Nspa$ is the maximum number of generated sparks. In this way, the worst fireworks in a population always receive at least $Nspa \cdot a$ sparks and the best fireworks get the maximum of $Nspa \cdot b$ sparks. It is implemented as follows:

$$\hat{s}_{ch} = \begin{cases} \mathrm{round}\left(a \cdot Nspa\right) & \text{for} \quad s_{ch} < a \cdot Nspa \\ \mathrm{round}\left(s_{ch}\right) & \text{for} \quad s_{ch} \in \left(a \cdot Nspa, b \cdot Nspa\right) \\ \mathrm{round}\left(b \cdot Nspa\right) & \text{for} \quad s_{ch} > b \cdot Nspa, \end{cases} \tag{6.10}$$

where $\mathrm{round}\left(\cdot\right)$ is a function approximating the real value of the argument to the nearest integer. After this operation, the total number of sparks generated by all fireworks is divided between the fireworks included in a given population.

Before sparks are generated it is necessary to determine the area of their location. The amplitude of explosion has to be determined for this purpose:

$$amp_{ch} = ampMax \cdot \frac{\mathrm{ff}\left(\mathbf{X}_{ch}\right) - \min\left\{\mathrm{ff}\left(\mathbf{X}_1\right), \ldots, \mathrm{ff}\left(\mathbf{X}_{Npop}\right)\right\} + \xi}{\sum_{i=1}^{Npop}\left(\mathrm{ff}\left(\mathbf{X}_i\right) - \min\left\{\mathrm{ff}\left(\mathbf{X}_1\right), \ldots, \mathrm{ff}\left(\mathbf{X}_{Npop}\right)\right\}\right) + \xi}, \tag{6.11}$$

where $ampMax > 0$ is the algorithm parameter indicating the maximum amplitude of explosion. The rule for determining the amplitude of explosion says that its value is inversely proportional to the value of the evaluation function. A high amplitude means that "good" individuals generate sparks in their surroundings, and vice versa. To prevent the concentration of sparks in too close proximity to the fireworks which are distinguished by a high value of the evaluation function, an additional restriction is applied [65]:

$$amp_{ch} := \begin{cases} ampMin \text{ if } amp_{ch} < ampMin \\ amp_{ch} \text{ if } \quad otherwise, \end{cases} \tag{6.12}$$

The value of the variable *ampMin* in the formula (6.12) can be systematically changed during the course of the algorithm operation. This is usually implemented as follows:

$$ampMin = ampInit - \frac{ampInit - ampFinal}{Niter} \cdot t, \tag{6.13}$$

or

$$ampMin = ampInit - \frac{ampInit - ampFinal}{Niter} \cdot \sqrt{(2 \cdot Niter - t) \cdot t}, \tag{6.14}$$

where t is the current step of the algorithm, *Niter* is the allowed number of steps of the algorithm, the parameter $ampFinal > 0$ is final, the minimum amplitude of explosion, the parameter $ampInit > ampFinal$ is initial minimum amplitude of explosion. A nonlinear function of the form (6.14) provides faster reduction of the explosion amplitude than a linear function of the form (6.13). The values of parameters *ampFinal* and *ampInit* can be made dependent on the value of the parameter *ampMax*. In the modification of the firework algorithm proposed in paper [65] t is the number of previous calls of the evaluation function and *Niter* is the maximum number of calls of the evaluation function. As previously mentioned, the number of calls of the evaluation function can also be a base for construction of the stopping criterion.

When the determination of the number of sparks and their explosion amplitude is completed, directions of sparks propagation should be determined. It is assumed that sparks for each firework are spread in a number of directions determined individually for each $\mathbf{X}_{ch}^{par}$. It can be lower or equal to L^{par}. This is conducted in such way that for each gene $X_{ch,g}^{par}$ of the individual $\mathbf{X}_{ch}^{par}$, the value round $\left(L^{par} \cdot U_g(0, 1)\right)$ is determined (function $U_g(0, 1)$ returns a real random number from the unit range). If the determined value is equal to 1, then the gene $X_{ch,g}^{par}$ of the individual $\mathbf{X}_{ch}^{par}$ associated with it is modified in accordance with the following dependency:

$$X_{ch,g}^{par} := X_{ch,g}^{par} + amp_{ch} \cdot U_{ch}(-1, 1), \tag{6.15}$$

where $\mathrm{U}_{ch}(-1, 1)$ is a random number from the range $[-1, 1]$, generated independently for each individual $\mathbf{X}_{ch}^{\text{par}}$. If, as a result of using the formula (6.15), the value of the gene $X_{ch,g}^{\text{par}}$ is outside the acceptable range $\left[\underline{X}_g^{\text{par}}, \overline{X}_g^{\text{par}}\right]$, then it is repaired in accordance with the following formula:

$$X_{ch,g}^{\text{par}} := \underline{X}_g^{\text{par}} + \left|X_{ch,g}^{\text{par}}\right| \,\%\, \left(\overline{X}_g^{\text{par}} - \underline{X}_g^{\text{par}}\right), \tag{6.16}$$

where $\underline{X}_g^{\text{par}}$ is the minimum allowed gene value, $\overline{X}_g^{\text{par}}$ is the maximum allowed gene value and $\%$ is the modulo operator. The values $\underline{X}_g^{\text{par}}$ and $\overline{X}_g^{\text{par}}$ result from the specific character of the considered problem.

Actualization of genes of the individual $\mathbf{X}_{ch}^{\text{par}}$, compatible with the dependencies (6.15) and (6.16), is called "generation of sparks".

In the modification of the firework algorithm proposed in paper [65] it is suggested that the procedure of sparks generation be streamlined. This is related to a change of the dependency (6.15) used in order to update the genes of the individual $\mathbf{X}_{ch}^{\text{par}}$:

$$X_{ch,g}^{\text{par}} := X_{ch,g}^{\text{par}} + amp_{ch} \cdot \mathrm{U}_g(-1, 1), \tag{6.17}$$

where $\mathrm{U}_g(-1, 1)$ is the random number from the range $[-1, 1]$, generated independently for each gene $X_{ch,g}^{\text{par}}$ of the individual $\mathbf{X}_{ch}^{\text{par}}$. Moreover, the modification of the firework algorithm involves changing the formula (6.16) used to repair gene $X_{ch,g}^{\text{par}}$, which, as a result of using the formula (6.17), is outside the acceptable range $\left[\underline{X}_g^{\text{par}}, \overline{X}_g^{\text{par}}\right]$. It has the following form:

$$X_{ch,g}^{\text{par}} := \underline{X}_g^{\text{par}} + \left(\overline{X}_g^{\text{par}} - \underline{X}_g^{\text{par}}\right) \cdot \mathrm{U}_g(0, 1), \tag{6.18}$$

where $\mathrm{U}_g(0, 1)$ is a random number from the range $[0, 1]$ generated independently for each gene $X_{ch,g}^{\text{par}}$ of the individual $\mathbf{X}_{ch}^{\text{par}}$.

Generated sparks are assessed using the defined evaluation function of the form (6.7).

Step 4. Generation of Additional Sparks

Generating additional random sparks in order to explore the search space involves random selection of *Nsparnd* fireworks from the fireworks set *Npop*. Then, the part encoding the parameters $\mathbf{X}_{ch}^{\text{par}}$ and the part encoding the structure $\mathbf{X}_{ch}^{\text{str}}$ are modified for each of the selected *Nsparnd* fireworks. Modification of the part $\mathbf{X}_{ch}^{\text{par}}$ encoding the parameters starts with determination of the directions of sparks propagation. This is similar to the previous step but the procedure for updating the selected genes of the individual $\mathbf{X}_{ch}^{\text{par}}$ is different. The revision is performed as follows:

$$X_{ch,g}^{\text{par}} := X_{ch,g}^{\text{par}} \cdot \mathrm{UGaussian}_{ch}(1, 1), \tag{6.19}$$

where $\text{UGaussian}_{ch}(1, 1)$ is a real number drawn from the Gaussian distribution (normal distribution). If the value of the gene $X^{\text{par}}_{ch,g}$ resulting from the application of the formula (6.19) is outside the acceptable range $\left[\underline{X}^{\text{par}}_{g}, \overline{X}^{\text{par}}_{g}\right]$, then it is repaired in accordance with the formula (6.16).

Modification of the firework algorithm [65] considered in the previous step also applies to the random sparks generation procedure. For randomly selected individual genes $\mathbf{X}^{\text{par}}_{ch}$ it proceeds as follows:

$$X^{\text{par}}_{ch,g} := X^{\text{par}}_{ch,g} + \left(XBest^{\text{par}}_{g} - X^{\text{par}}_{ch,g}\right) \cdot \text{UGaussian}_{ch}(0, 1), \tag{6.20}$$

where $\text{UGaussian}_{ch}(0, 1)$ is a real number drawn from the Gaussian distribution (normal distribution), $XBest^{\text{par}}_{g}$ is the firework with currently the best value of the evaluation function. If the value of gene $X^{\text{par}}_{ch,g}$ resulting from the application of the formula (6.20) is outside the acceptable range $\left[\underline{X}^{\text{par}}_{g}, \overline{X}^{\text{par}}_{g}\right]$, then it is repaired in accordance with the formula (6.18).

In addition to the genes encoding parameters, the genes encoding the structure are also modified. Modification of the selected individuals $\mathbf{X}^{\text{str}}_{ch}$ encoding the structure is performed using the mutation operator typical for the genetic algorithm. Therefore, for each gene of the modified individuals $\mathbf{X}^{\text{str}}_{ch}$ a random number from the unit range is drawn. If it is lower than the value of so-called mutation probability $p_m \in (0, 1)$ (which is a parameter of the algorithm), then the value of the gene is changed to the opposite (i.e. from 0 to 1 and vice versa).

Additional sparks generated in this step are assessed using the defined evaluation function (6.7).

Step 5. Creation of a New Population

The new population of individuals is joined by the currently best firework (indicated by the index *chBest* and having the lowest value of the minimized fitness function of the form (6.7)) and *Npop* − 1 individuals selected, e.g. using the roulette wheel method [55], from the sparks and other fireworks generated in steps 3 and 4. Selection probability of the individual $\mathbf{X}_{ch}$ is determined for the roulette wheel method as follows:

$$p\left(\mathbf{X}_{ch}\right) = \frac{\sum\limits_{\substack{ch2=1,\\ ch2\neq chBest}}^{Npop+Nspa} \left\|\mathbf{X}_{ch} - \mathbf{X}_{ch2}\right\|}{\sum\limits_{\substack{ch3=1,\\ ch3\neq chBest}}^{Npop+Nspa} \sum\limits_{\substack{ch2=1,\\ ch2\neq chBest}}^{Npop+Nspa} \left\|\mathbf{X}_{ch3} - \mathbf{X}_{ch2}\right\|}, \tag{6.21}$$

where $\|\cdot\|$ is an adopted distance measure (e.g. Euclidean). In the roulette wheel method each individual which is a candidate to a new population is assigned to a segment of the wheel whose size is proportional to the probability of the selection of the individual $p\left(\mathbf{X}_{ch}\right) \cdot 100\%$. In this way the virtual roulette wheel is obtained.

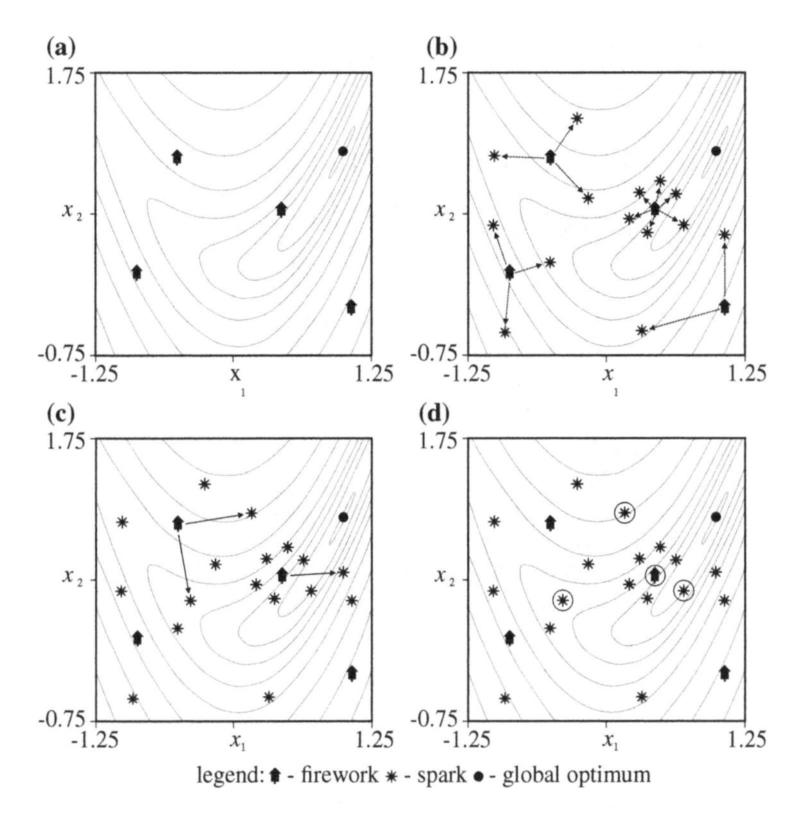

Fig. 6.2 Illustration of selected aspects of the operation of the hybrid genetic-firework algorithm in a two-dimensional space, shown for the Rosenbrock function (determined by the formula $f(\mathbf{x}) = (1-x_1)^2 + 10 \cdot (x_2 - x_1^2)^2$, with the minimum in the point $x_1 = 1, x_2 = 1$): **a** initialization of a fireworks population, **b** generation of sparks for fireworks, **c** generation of additional sparks, **d** selection of individuals (denoted as *circles*) creating a new fireworks population

Then, a single number from the range [0, 100] which indicates the individual selected to the new population is drawn.

In practice other selection methods can also be used. The suggested method of selection for modification of the firework algorithm described in [65] is Elitism-Random Selection [25].

Step 6. Replacement of the Population

In this step the old population of individuals is replaced by the population generated in the previous step. In the new population all individuals are treated as fireworks.

A sample illustration of the selected aspects of the hybrid genetic-firework algorithm is shown in Fig. 6.2.

6.2.2 *Evolutionary Learning of Fuzzy Systems Using Many Populations*

The previous section describes a sample hybrid one-population algorithm dedicated to evolutionary learning of fuzzy systems. In this section we focus on a hybrid multi-population algorithm, which was created by combining the genetic algorithm [34] and the imperialist algorithm [6]. It will be denoted as the GIA. The genetic algorithm is meant to select a fuzzy system structure while the applied imperialist algorithm is meant to select fuzzy system parameters. The idea of the imperialist algorithm is inspired by the social evolution whereas the idea of the genetic algorithm is inspired by the biological evolution of species. Each individual of a population in the terminology of the imperialist algorithm is a "country". "Imperialists" are selected from these countries. They are "countries" with the best value of the evaluation function (i.e. the lowest possible value in the case of the considered problem of minimizing the evaluation function (6.7)). "Imperialists" create "empires", for which other "countries" are "colonies". "Colonies" are subject of evolutionary operations referring to the social evolution. These operations are assimilation and revolution. In addition, a binary mutation of the "colony" is performed (part encoding structure). It is derived from the genetic algorithm. Evolutionary operations change "the balance of power" of given "empires". After this change, a decision is made whether the process of the social evolution should continue, or whether the solution (a fuzzy system encoded in an individual of a population) meeting the stopping criterion of the algorithm has been obtained. The purpose of the algorithm is a systematic improvement of solutions in the sense of the adopted evaluation function, which is implemented in the following steps of the algorithm. These steps are shown in Fig. 6.3. The considered algorithm uses encoding presented in Sect. 6.2.1.1 and the way of evaluating of individuals presented in Sect. 6.2.1.2.

6.2.2.1 Processing of a Population

The hybrid genetic-imperial algorithm works as shown in the block diagram presented in Fig. 6.3. The following part of this section presents a detailed description of the steps performed by the algorithm.

Step 1. Initialization of the Algorithm

Initialization of a population is performed before the start of the algorithm. In the basic initialization procedure of the population the genes of all individuals (countries) $\mathbf{X}_{ch}$, $ch = 1, \ldots, Npop$ ($Npop$ is the number of countries) can be initialized randomly. Thus, binary values of the genes of the part $\mathbf{X}_{ch}^{\text{str}}$ encoding fuzzy system structure can be drawn from the set $X_{j,g}^{\text{str}} \in \{0, 1\}$, where the gene index $g = 1, \ldots, L^{\text{str}}$. Value 1 of the gene means that the connection associated with this gene occurs in the fuzzy system structure. Real values of genes of the part $\mathbf{X}_{ch}^{\text{par}}$ encoding fuzzy system parameters can be drawn from the range determined individually for the problem

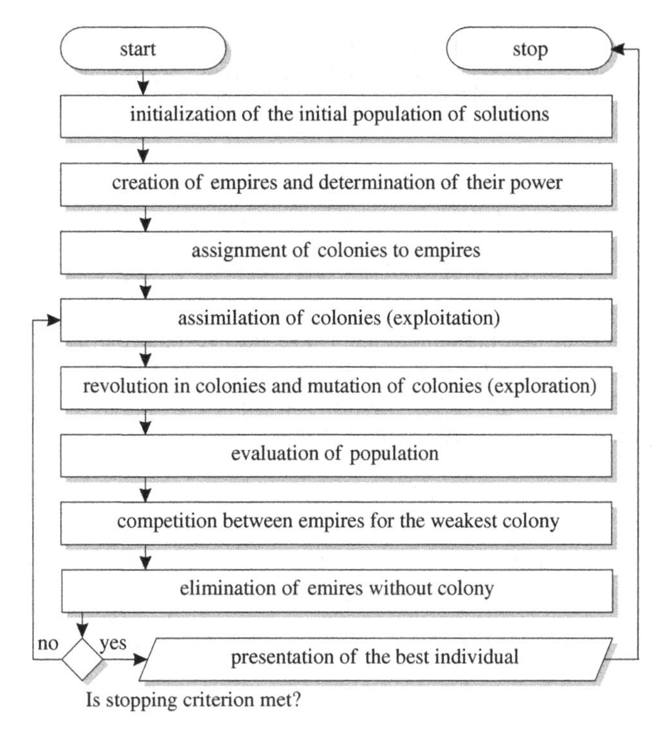

Fig. 6.3 Block scheme of the hybrid genetic-imperialist algorithm for automatic selection of structure and parameters of a fuzzy system

under consideration ($X^{\text{par}}_{ch,g} \in \left[\underline{X}^{\text{par}}_{g}, \overline{X}^{\text{par}}_{g}\right]$). In the initialization procedure we can also use other initialization methods, e.g. those mentioned in Sect. 6.1.

Step 2. Creation of Empires and Determination of Their Power

In the procedure used for creation of empires all individuals $\mathbf{X}_j$ of a given population are divided on the basis of the value of the evaluation function into two groups: countries and empires. The empire group (denoted as $\mathbf{Xi}_k = \left\{\mathbf{Xi}^{\text{str}}_k, \mathbf{Xi}^{\text{par}}_k\right\}$ $= \left\{Xi_{k,1}, \ldots, Xi_{k,L^{\text{str}}+L^{\text{par}}}\right\}$) contains those individuals of the population ($k = 1, \ldots, Ni$) which have the best value of the evaluation function. The number of empires Ni can be set freely, but in the literature a suggestion has been found that $Ni = \text{round}\left(\frac{N}{10}\right)$ (round $(\cdot)$ is a function approximating to the nearest integer) [6]. The colony group comprises all the other individuals of the population. Their number is equal to $N - Ni$.

After the division of the population into empires and colonies, colonies are assigned to empires. This process takes into account the power of empires P_k $(k = 1, \ldots, Ni)$ determined as follows:

$$Pi_k = \left| \frac{\mathrm{ff}\left(\mathbf{Xi}_k\right) - \max\limits_{s=1,\ldots,Ni} \left\{\mathrm{ff}\left(\mathbf{Xi}_s\right)\right\}}{\sum\limits_{q=1}^{Ni} \left(\mathrm{ff}\left(\mathbf{Xi}_q\right) - \max\limits_{s=1,\ldots,Ni} \left\{\mathrm{ff}\left(\mathbf{Xi}_s\right)\right\}\right)} \right|, \tag{6.22}$$

where the numerator and the denominator take into account the normalized value of the evaluation function for the problem of minimizing the value of the evaluation function. $Nic_k = \mathrm{int}\left(Ni \cdot Pi_k\right)$ colonies are randomly assigned to each empire. Colonies assigned to the empire k will be denoted as:

$$\mathbf{Xic}_{k,r} = \left\{\mathbf{Xic}_{k,r}^{\mathrm{str}}, \mathbf{Xic}_{k,r}^{\mathrm{par}}\right\} = \left\{Xic_{k,r,1}, \ldots, Xic_{k,r,L^{\mathrm{str}}+L^{\mathrm{par}}}\right\}, \tag{6.23}$$

where $k = 1, \ldots, Ni$ and $r = 1, \ldots, Nic_k$. The system of empires and their colonies created in this step is subject to changes described in the subsequent steps.

Step 3. Changes in Colonies

The purpose of making changes in colonies is exploitation and exploration of the search space of considerations. Exploitation is meant to make colonies similar to empires (as individuals with the best values of the evaluation function) for which they have been assigned. This is realized using an assimilation operator typical of the imperialist algorithm. On the other hand, exploration is meant to introduce random changes in the population allowing for finding unknown good solutions. This is implemented using a revolution operator typical of the imperialist algorithm, and a mutation operator typical of the genetic algorithm.

The assimilation operator processes the part $\mathbf{Xc}_r^{\mathrm{par}}$ of individuals $\mathbf{Xc}_r$, which encode fuzzy system parameters. It allows colonies to move toward empires, but at the same time it introduces a small offset by a randomly selected angle. Its operation can be defined as follows:

$$Xic_{k,r,g}^{\mathrm{par}} := \left(Xi_{k,g}^{\mathrm{par}} - Xic_{k,r,g}^{\mathrm{par}}\right) \cdot \mathrm{U}_r\left(0, 2\right) \cdot \mathrm{U}_g\left(-\gamma, \gamma\right), \tag{6.24}$$

where $\mathrm{U}_r\left(0, 2\right)$ is a random number from the range $(0, 2)$ generated for the purposes of assimilation of the colony r, $\mathrm{U}_g\left(-\gamma, \gamma\right)$ is a random number from the range $(-\gamma, \gamma)$ generated individually for each gene g of the colony r, $\gamma > 0$ is a parameter of the algorithm determining intensity of the random offset. The assimilation operator allows empires to keep their strong position as good solutions in the sense of the evaluation function value. It also prevents the algorithm from making those changes in individuals of the population which would make it impossible to find the optimal solution in terms of the adopted criteria.

After the assimilation process, revolution and mutation of the colonies are performed in the population. In contrast to the assimilation, revolution and mutation only affect a subset of the genes of the colony. It is implemented in such way that for each gene of each colony a real number from the range [0, 1] is drawn. If the number value is lower than so-called revolution and mutation probability $p_{rm} \in [0, 1]$ (being parameter of the algorithm and working in a similar way to the probability of mutation in genetic algorithms), then this gene is subject to evolution or mutation. Revolution mainly refers to the genes encoding fuzzy system parameters ($\mathbf{Xc}_r^{\text{par}}$) while mutation, on the other hand, refers to the genes encoding the fuzzy system structure ($\mathbf{Xc}_r^{\text{str}}$). The operation of the revolution operator can be expressed as follows:

$$Xic_{k,r,g}^{\text{par}} := \underline{X}_g^{\text{par}} + \left(\overline{X}_g^{\text{par}} - \underline{X}_g^{\text{par}}\right) \cdot \text{U}_g\left(0, 1\right), \tag{6.25}$$

where $\underline{X}_g^{\text{par}}$ is the minimum allowed value of the gene, $\overline{X}_g^{\text{par}}$ is the maximum allowed value of the gene (similar as in the GFA algorithm).

On the other hand, operation of the mutation operator consists in changing the value of the binary gene encoding the information on using (or blocking) of the associated component of a fuzzy system to the opposite information (i.e. from 1 to 0 or vice versa). Since the revolution and mutation affect colonies significantly, the value of parameter p_{rm} cannot be too large so as not to cause degeneration of the population.

Step 4. Changes in Empires

Changes in empires are implemented in two stages. The first one involves competition of each empire $\mathbf{Xi}_k$ with its best colony ($\mathbf{Xic}_{k,r}$). The base of this competition is the value of the evaluation function. The second stage involves moving the weakest colony of the weakest empire to another empire. The process of determining of the weakest empire takes into account the total power of empires and the probability of the weakest colony acquisition, which is determined on the basis of that power. The total power of empires is determined as follows:

$$C_k = \text{ff}\left(\mathbf{Xi}_k\right) + \zeta \cdot \frac{\sum_{r=1}^{Nci_k} \text{ff}\left(\mathbf{Xic}_{k,r}\right)}{Nci_k}, \tag{6.26}$$

is $\zeta \in [0, 1]$ is a coefficient of colony importance (algorithm parameter). With the total power of empires having been computed, the probability of acquiring of the weakest colony is determined for each empire (similar as in formula (6.22)):

$$Pic_k = \left| \frac{C_k - \max_{s=1,\ldots,Ni} \{C_s\}}{\sum_{q=1}^{Ni} \left(C_q - \max_{s=1,\ldots,Ni} \{C_s\} \right)} \right|. \tag{6.27}$$

The sum of probabilities of the weakest colony acquisition for all empires is equal to 1. For the most powerful empire this probability is the highest and for the weakest one it equals 0. To indicate the empire which will acquire the weakest colony, we can use the roulette wheel method. Then, each empire has to be associated with a segment of the wheel, whose size is proportional to the probability of acquiring the worst colony Pic_k. The virtual roulette wheel prepared in this way can be used in the selection process similar as in Sect. 6.2.1 or in the genetic algorithm [55].

Moving the weakest colonies between empires eliminates the weakest empires in subsequent steps of the algorithm. This is implemented in such way that the empires that have no colonies are removed. Moreover, movement of the weakest colonies between empires results in weakening of the strongest empires. As a result, the algorithm becomes less sensitive to the local minima.

A sample illustration of the selected aspects of the hybrid genetic-imperialist algorithm is shown in Fig. 6.4.

6.3 Taking into Account Interpretability Criteria of Fuzzy Systems

In Sects. 3.4 and 4.4 different interpretability criteria and their assumptions are discussed. This section contains a sample way of their implementation. It depends on the type of the applied membership function and the type of the aggregation and inference operators used. Due to this, the criteria described in this section have been defined for the Gaussian membership function (4.29) and the operators built on the basis of the parameterized Dombi-type triangular norms of the form (4.1) (similar as in Sect. 6.2)). The criteria considered in this section have been designed in such a way so as to be minimized and aggregated by the function of the form (6.8).

It is worth noting that in the approach based on the sets of values (mentioned in Sect. 4.4), the use of interpretability criteria can be omitted. However, it is not always possible to determine the values included in these sets.

6.3.1 Complexity Evaluation Criterion

The complexity evaluation criterion of the fuzzy system encoded in an individual $\mathbf{X}_{ch}$ of the form (6.1) can be defined as follows:

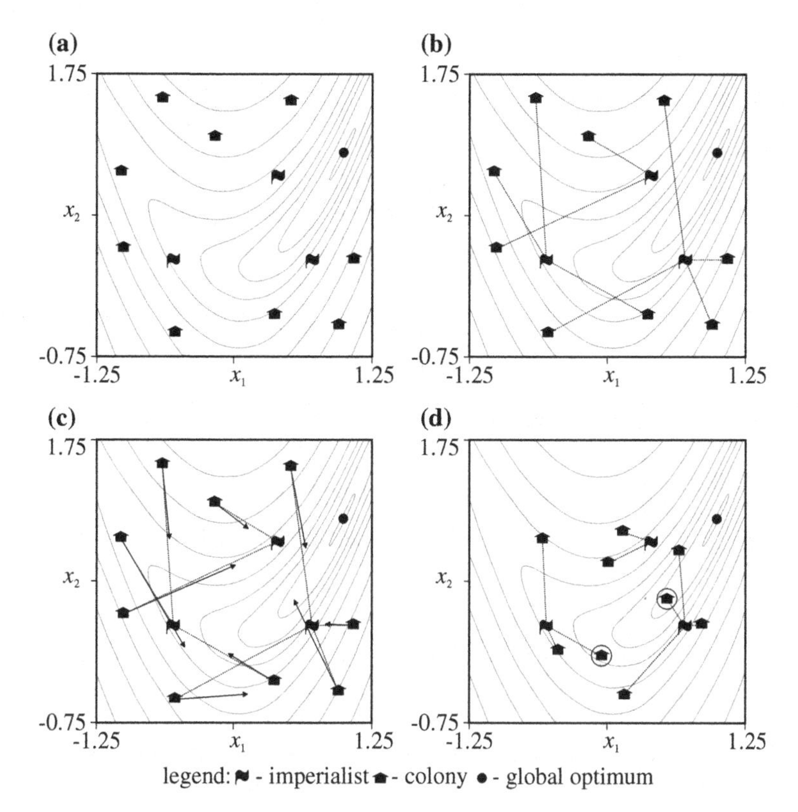

Fig. 6.4 Illustration of the selected aspects of the operation of the hybrid genetic-imperialist algorithm in a two-dimensional space shown for the Rosenbrock function (determined by the formula $f(\mathbf{x}) = (1 - x_1)^2 + 10 \cdot \left(x_2 - x_1^2\right)^2$, with the minimum in point $x_1 = 1, x_2 = 1$): **a** division of the population into imperialists and colonies, **b** creation of empires, **c** assimilation and revolution, **d** acquisition of empires by colonies with a better value of the evaluation function (denoted in *circles*)

$$\text{ffint}_A\left(\mathbf{X}_{ch}\right) = \frac{\text{noifs}\left(ch\right) + \text{noofs}\left(ch\right) + \text{nod}\left(ch\right)}{Nmax \cdot (n+m) + Rmax \cdot m}, \tag{6.28}$$

where the functions $\{\text{noifs}\left(\cdot\right), \text{noofs}\left(\cdot\right), \text{nod}\left(\cdot\right)\}$ have been defined in Table 6.1.

6.3.2 *Fuzzy Sets Position Evaluation Criterion*

The fuzzy set placement evaluation criterion of the fuzzy system encoded in the individual $\mathbf{X}_{ch}$ of the form (6.1) can be defined as follows:

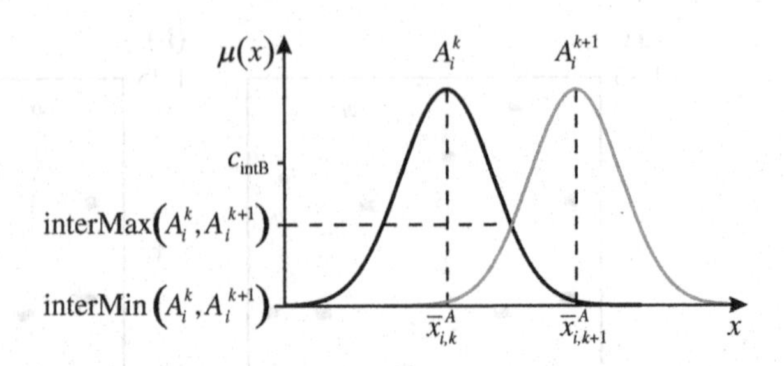

Fig. 6.5 Graphical interpretation of the intersection points of fuzzy sets and the desired intersection points of fuzzy sets

$$\text{fint}_B(\mathbf{X}_{ch}) = \frac{1}{2}\left(\begin{array}{c} \dfrac{\frac{1}{2}\cdot\sum\limits_{i=1}^{\text{noi}(ch)}\sum\limits_{k=1}^{\text{noifsfi}(ch,i)-1}\left(\begin{array}{c}\left|c_{\text{intB}} - \text{interMax}\left(A_i^k, A_i^{k+1}\right)\right| + \\ +\text{interMin}\left(A_i^k, A_i^{k+1}\right)\end{array}\right)}{\sum\limits_{i=1}^{\text{noi}(ch)}(\text{noifsfi}(ch,i)-1)} + \\ + \dfrac{\frac{1}{2}\cdot\sum\limits_{j=1}^{m}\sum\limits_{k=1}^{\text{noofsfo}(ch,j)-1}\left(\begin{array}{c}\left|c_{\text{intB}} - \text{interMax}\left(B_j^k, B_j^{k+1}\right)\right| + \\ +\text{interMin}\left(B_j^k, B_j^{k+1}\right)\end{array}\right)}{\sum\limits_{j=1}^{m}(\text{noofsfo}(ch,j)-1)} \end{array}\right), \quad (6.29)$$

where $c_{\text{intB}} \in (0, 1)$ determines the desired value of the membership function at the intersection point between two adjacent fuzzy sets (in the simulations we assumed the value 0.5), the functions $\{\text{noi}(\cdot), \text{noifsfi}(\cdot), \text{noofsfo}(\cdot)\}$ are defined in Table 6.1, the functions $\{\text{interMax}(\cdot), \text{interMin}(\cdot)\}$ determine the values of two intersection points of the fuzzy sets (Fig. 6.5). For the fuzzy sets expressed by the Gaussian function of the form (4.29) they take the following form:

$$\begin{cases} \text{interMin}\left(A_i^k, A_i^{k+1}\right) = \exp\left(-\dfrac{\left(\mathbf{X}_{ch}^{\text{supp}}\{\bar{x}_{i,k}^A\} - \mathbf{X}_{ch}^{\text{supp}}\{\bar{x}_{i,k+1}^A\}\right)^2}{\left(\mathbf{X}_{ch}^{\text{supp}}\{\sigma_{i,k}^A\} - \mathbf{X}_{ch}^{\text{supp}}\{\sigma_{i,k+1}^A\}\right)^2}\right), \\ \text{interMax}\left(A_i^k, A_i^{k+1}\right) = \exp\left(-\dfrac{\left(\mathbf{X}_{ch}^{\text{supp}}\{\bar{x}_{i,k}^A\} - \mathbf{X}_{ch}^{\text{supp}}\{\bar{x}_{i,k+1}^A\}\right)^2}{\left(\mathbf{X}_{ch}^{\text{supp}}\{\sigma_{i,k}^A\} + \mathbf{X}_{ch}^{\text{supp}}\{\sigma_{i,k+1}^A\}\right)^2}\right), \end{cases} \quad (6.30)$$

where $\mathbf{X}_{ch}^{\text{supp}}$ is a temporary set of the fuzzy system parameters containing the parameters of the input and output fuzzy sets sorted ascending in relation to the centers of these sets:

$$\mathbf{X}_{ch}^{\text{supp}} = \left\{\begin{array}{c} \bar{x}_{1,1}^A, \sigma_{1,1}^A, \ldots, \bar{x}_{1,\text{noifsfi}(ch,1)}^A, \sigma_{1,\text{noifsfi}(ch,1)}^A, \ldots, \\ \bar{x}_{\text{noi}(ch),1}^A, \sigma_{\text{noi}(ch),1}^A, \ldots, \bar{x}_{\text{noi}(ch),\text{noifsfi}(ch,\text{noi}(ch))}^A, \sigma_{\text{noi}(ch),\text{noifsfi}(ch,\text{noi}(ch))}^A, \\ \bar{y}_{1,1}^B, \sigma_{1,1}^B, \ldots, \bar{y}_{1,\text{noofsfo}(ch,1)}^B, \sigma_{1,\text{noofsfo}(ch,1)}^B, \ldots, \\ \bar{y}_{m,1}^B, \sigma_{m,1}^B, \ldots, \bar{y}_{m,\text{noofsfo}(ch,m)}^B, \sigma_{m,\text{noofsfo}(ch,m)}^B \end{array}\right\}. \quad (6.31)$$

The criterion (6.29) takes values close to 0 when values of the function $\mathrm{interMax}(\cdot) \approx c_{\mathrm{intB}}$ and values of the function $\mathrm{interMin}(\cdot) \approx 0$ (Fig. 6.5). In this case, the position of fuzzy sets can be considered as readable.

6.3.3 *Criterion for Assessing Similarity of Fuzzy Sets Width*

The criterion for assessing similarity of fuzzy sets width of the fuzzy system encoded in the individual $\mathbf{X}_{ch}$ of the form (6.1) can be defined as follows:

$$\mathrm{ffint}_C(\mathbf{X}_{ch}) = \frac{1}{2} \cdot \left(\frac{\sum_{i=1}^{\mathrm{noi}(ch)} \sum_{k1=1}^{\mathrm{noifsfi}(ch,i)-1} \sum_{k2=k1+1}^{\mathrm{noifsfi}(ch,i)} \mathrm{sim}_{width}\left(A_i^{k1}, A_i^{k2}\right)}{\sum_{i=1}^{\mathrm{noi}(ch)} \binom{\mathrm{noifsfi}(ch,i)}{2}} + \frac{\sum_{j=1}^{m} \sum_{k1=1}^{\mathrm{noofsfo}(ch,j)} \sum_{k2=k1+1}^{\mathrm{noofsfo}(ch,j)-1} \mathrm{sim}_{width}\left(B_j^{k1}, B_j^{k2}\right)}{\sum_{j=1}^{m} \binom{\mathrm{noofsfo}(ch,j)}{2}} \right), \tag{6.32}$$

where $\binom{n}{k}$ is the Newton's binomial, the function $\mathrm{sim}_{width}(\cdot)$ is used to determine similarity of width of the membership function of fuzzy sets. For the sets expressed by the Gaussian function (4.29) it may take the following form:

$$\mathrm{sim}_{width}\left(A_i^{k1}, A_i^{k2}\right) = \frac{\left|\mathbf{X}_{ch}^{\mathrm{supp}}\left\{\sigma_{i,k1}^{A}\right\} - \mathbf{X}_{ch}^{\mathrm{supp}}\left\{\sigma_{i,k2}^{A}\right\}\right|}{\max\left\{\mathbf{X}_{ch}^{\mathrm{supp}}\left\{\sigma_{i,k1}^{A}\right\}, \mathbf{X}_{ch}^{\mathrm{supp}}\left\{\sigma_{i,k2}^{A}\right\}\right\}}, \tag{6.33}$$

where $\mathbf{X}_{ch}^{\mathrm{supp}}$ is a temporary set of the system parameters of the form (6.31). A sample graph of the function $\mathrm{sim}_{width}(\cdot)$ is shown in Fig. 6.6.

The criterion (6.32) takes values close to 0 when fuzzy sets have a similar width.

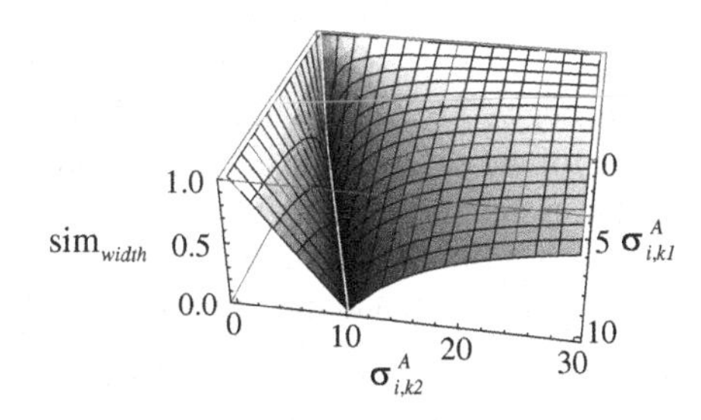

Fig. 6.6 Sample graph of the function $\mathrm{sim}_{width}(\cdot)$ for $\sigma_{i,k1}^{A} \in [0, 10]$ i $\sigma_{i,k2}^{A} \in [0, 30]$

6.3.4 Criterion for Assessing Coverage of the Data Space by Input Fuzzy Sets

The criterion for assessing coverage of the data space by input fuzzy sets of the fuzzy system encoded in the individual $\mathbf{X}_{ch}$ of the form (6.1) can be defined as follows:

$$\text{ffint}_D\left(\mathbf{X}_{ch}\right) = \frac{\sum_{z=1}^{Z}\sum_{i=1}^{n}\left(\mathbf{X}_{ch}^{\text{str}}\left\{x_i\right\}\cdot\max\left\{0, \text{neg}_Z\left(\sum_{k=1}^{Nmax}\left(\begin{array}{c}\mathbf{X}_{ch}^{\text{str}}\left\{rule_k\right\}\cdot\\ \cdot\mu_{A_i^k}\left(\bar{x}_{z,i}^L\right)\end{array}\right)\right)\right\}\right)}{Z\cdot\text{noi}\left(ch\right)}, \tag{6.34}$$

where $\text{neg}_Z\left(\cdot\right)$ is Zadeh's negation of the form (2.14), $\bar{x}_{z,i}^L$ is the value of input signal i from the set z of learning sequence.

Criterion (6.34) takes values close to 0 when for each set from the learning sequence the sum of its membership in input fuzzy sets is equal to 1 or more. In this case uniformity of coverage of the data space by input fuzzy sets can be regarded as satisfactory. The criterion of the form (6.34) is complementary to the criterion of the form (6.32) related to similarity of fuzzy sets width and the criterion (6.29) related to the readability of fuzzy sets position.

6.3.5 Criterion for Assessing Fuzzy Rules Activity

The criterion for assessing fuzzy rules activity of the fuzzy system encoded in the individual $\mathbf{X}_{ch}$ of the form (6.1) can be defined as follows:

$$\text{ffint}_E\left(\mathbf{X}_{ch}\right) = \text{neg}_Z\left(\frac{1}{Z}\cdot\sum_{z=1}^{Z}\frac{\left(\max_{k=1,\ldots,Nmax}\left\{\mathbf{X}_{ch}^{\text{str}}\left\{rule_k\right\}\cdot\tau_k\left(\bar{\mathbf{x}}_z\right)\right\}\right)}{\sum_{k=1}^{Nmax}\mathbf{X}_{ch}^{\text{str}}\left\{rule_k\right\}\cdot\tau_k\left(\bar{\mathbf{x}}_z\right)}\right), \tag{6.35}$$

where $\text{neg}_Z\left(\cdot\right)$ is Zadeh's negation of the form (2.14), $\tau_k\left(\cdot\right)$ is the activity level of the rule k for the set from the learning sequence z (e.g. expressed by the formula (4.27) for the system of the form (4.28)).

The criterion (6.35) takes values close to 0 when the activity level of one of the rules is close to 1, and the activity level of the other rules is close to 0.

6.3.6 *Criterion for Assessing the Readability of Weight Values in the Fuzzy Rule Base*

The criterion for assessing the readability of weight values in the fuzzy rule base of the fuzzy system encoded in the individual $\mathbf{X}_{ch}$ of the form (6.1) can be defined as follows:

$$\text{ffint}_F(\mathbf{X}_{ch}) = \\ = \text{neg}_Z\left(\frac{\sum_{k=1}^{Nmax}\mathbf{X}_{ch}^{\text{str}}\{rule_k\}\cdot\left(\begin{array}{c}\sum_{i=1}^{n}\mathbf{X}_{ch}^{\text{str}}\{x_i\}\cdot\mathbf{X}_{ch}^{\text{str}}\{A_i^k\}\cdot\mu_w\left(w_{i,k}^A\right)+\\ +\sum_{j=1}^{m}\mathbf{X}_{ch}^{\text{str}}\left\{B_j^k\right\}\cdot\mu_w\left(w_{i,k}^B\right)+\\ +\sum_{k=1}^{Nmax}\mu_w\left(w_k^{\text{rule}}\right)\end{array}\right)}{\text{noifs}(ch)+\text{noofs}(ch)+\text{nor}(ch)}\right), \tag{6.36}$$

where $\text{neg}_Z(\cdot)$ is Zadeh's negation of the form (2.14), $\mu_w(\cdot)$ is a function "promoting" obtaining the values of 0.0, 0.5 and 1.0 expressed by the formula (4.30).

The criterion (6.36) takes values close to 0 when all weights take the values close to the values from the set {0.0, 0.5, 1.0}.

6.3.7 *Criterion for Assessing the Readability of Aggregation and Inference Operators*

The criterion for assessing the readability of aggregation and inference operators of the fuzzy system encoded in the individual $\mathbf{X}_{ch}$ of the form (6.1) can be defined as follows (Fig. 4.3):

$$\text{ffint}_G(\mathbf{X}_{ch}) = \text{neg}_Z\left(\frac{1}{3}\cdot\left(\begin{array}{c}\mu_p\left(\mathbf{X}_{ch}^{\text{par}}\{p^\tau\}\right)+\\ +\mu_p\left(\mathbf{X}_{ch}^{\text{par}}\{p^{\text{imp}}\}\right)+\\ +\mu_p\left(\mathbf{X}_{ch}^{\text{par}}\{p^{\text{agr}}\}\right)\end{array}\right)\right), \tag{6.37}$$

where $\mu_p(\cdot)$ is a function "promoting" obtaining the expected values, adapted to the type of the triangular norms applied. Sample values for the Dombi-type triangular norms are presented in Table 4.3, and the function $\mu_p(\cdot)$ dedicated to them is described by the formula (4.31).

6.3.8 Criterion for Assessing the Defuzzification Mechanism

Defuzzification mechanism does not have a direct impact on interpretability of a fuzzy system. However, this criterion can be used to evaluate usefulness of discretization points of a given defuzzification procedure. Then, the evaluation criterion of the defuzzification mechanism of a fuzzy system encoded in the individual of the population $\mathbf{X}_{ch}$ of the form (6.1) can be defined as follows:

$$\text{ffint}_H\left(\mathbf{X}_{ch}\right) = \text{neg}_Z\left(\frac{\sum_{j=1}^{m}\sum_{r=1}^{Rmax}\left(\mathbf{X}_{ch}^{\text{str}}\left\{\bar{y}_{j,r}^{\text{def}}\right\}\cdot\max_{k=1,\ldots,Nmax}\left\{\begin{array}{c}\mathbf{X}_{ch}^{\text{str}}\left\{B_j^k\right\}\cdot\\ \cdot\mu_y\left(\mu_{B_j^k}\left(\bar{y}_{j,r}^{\text{def}}\right)\right)\end{array}\right\}\right)}{\sum_{j=1}^{m}\sum_{r=1}^{Rmax}\mathbf{X}_{ch}^{\text{str}}\left\{\bar{y}_{j,r}^{\text{def}}\right\}}\right), \tag{6.38}$$

where $\text{neg}_Z(\cdot)$ is Zadeh's negation of the form (2.14) and $\mu_y(\cdot)$ is the function used to control discretization points activity. It can have the following form:

$$\mu_y(x) = \begin{cases} 1 \text{ for } x \geq c_{\text{intH}} \\ 0 \text{ for } x < c_{\text{intH}}, \end{cases} \tag{6.39}$$

where $c_{\text{intH}} \in [0, 1]$ is the minimum expected value of the membership function for discretization points (the parameter used for evaluation of discretization points activity).

The criterion (6.38) determines the number of "active" discretization points in relation to the total number of discretization points. It takes values equal to 0 when all discretization points are "active". An "active" discretization point is the one for which the membership function of any output fuzzy set is greater than or equal to the constant c_{intH}.

In practice different criteria relating to defuzzification mechanism can be defined. They can verify not only inclusion of the output data in the subdomain, but also similarity to the characteristic points of this subdomain. Such points can be centers of gravity of output fuzzy sets, points of their intersection, fuzzy sets characteristic points B_j', etc.

6.4 Simulation Results

The remarks on the simulations can be summarized as follows:

- The aim of the simulations was to present examples of how interpretability of fuzzy systems designed by using evolutionary learning can be changed (Table 6.2).

- In the learning phase we used the hybrid genetic-firework algorithm (GFA) discussed in Sect. 6.2.1 and whose parameters are presented in Table 6.3, and the hybrid genetic-imperialist algorithm (GIA) outlined in Sect. 6.2.2 and whose parameters are presented in Table 6.4. Moreover, in the learning phase the interpretability criteria described in Sect. 6.3 were taken into account.
- The information on the used simulation problems is presented in Table 6.2. They are two sample approximation problems.
- In the simulation we used a Mamdani-type system of the form (4.28), whose parameters are presented in Table 6.5.
- The simulations were performed for seven different combinations of weights of the evaluation function (6.7): from the one aimed at accuracy (V1) to the one aimed at interpretability (V7). The set of these variants is shown in Table 6.6.
- The simulations were performed taking into account all the interpretability criteria described in Sect. 6.3. The function of the form (6.8) was used to aggregate these criteria. The following weight values were assigned to the criteria: $w_{\text{ffintA}} = 0.5$, $w_{\text{ffintB}} = 1.0$, $w_{\text{ffintC}} = 0.5$, $w_{\text{ffintD}} = 0.5$, $w_{\text{ffintE}} = 0.2$, $w_{\text{ffintF}} = 0.2$, $w_{\text{ffintG}} = 0.2$, $w_{\text{ffintH}} = 0.2$. The adopted values refer to the assumptions of the semantics outlined in Sect. 3.3, which was presented in the paper [3].
- Each simulation (for each variant, i.e. V1..V7) was repeated 100 times drawing a population of individuals of the form (6.1) every time. The obtained results were averaged and they are presented in Tables 6.7, 6.8 and 6.9 and in Figs. 6.7, 6.8, 6.9, 6.10, 6.11 and 6.12. Due to different complexity of the considered simulation problems, Figs. 6.11 and 6.12 are regarded as indicative.
- The fuzzy rules are presented in Tables 6.7 and 6.8 and in Figs. 6.7 and 6.9. Each fuzzy set and each fuzzy rule has an assigned weight of importance represented in the figures by a rectangle. Filling of the rectangle depends on the weight value, i.e. a complete filling means that the weight value is equal to 1.0, no filling means that the weight value equals 0.0. In the graphic representation of the fuzzy rules the values of the Dombi-type norms' parameters are also taken into account.
- The symbols of the weights in the notation of the rules of the form (4.24) were replaced by linguistic labels (Tables 6.7 and 6.8). They have the following form: '*v*' if a weight is greater than 0.75 (*very important*), '*i*' if the weight value is in the range [0.25, 0.75] (*important*), and '*n*' if the value of the weight is less than 0.25 (*not important*).
- The names of the input fuzzy sets A_i^k and output fuzzy sets B_j^k in the notation of rules in the form (4.24) were replaced by the following linguistic labels: '*very low*', '*low*', '*medium low*', '*medium*', '*medium high*', '*high*', '*very high*' (Tables 6.7 and 6.8). The fuzzy sets which had been reduced in the system were not taken into account in the notation of rules of the form (4.24). Sometimes in the literature these sets are labelled as '*don't care*'. In the case when a fuzzy system had only one fuzzy set assigned to a specified input or output, it was labelled as '*near* [value]'.
- The names of the inputs and outputs in the notation of fuzzy rules of the form (4.24) were replaced by linguistic variables taken from the description of the discussed simulation problems (Tables 6.7 and 6.8). Moreover, these names were extended by the index of an input or output placed in square brackets (e.g. '$\textit{frequency}_{[1]}$').

Table 6.2 A set of test problems used in the simulations of affecting interpretability in evolutionary learning

Item no.	Problem name	Number of input attributes	Number of output attributes	Number of sets	Problem type	Problem label
1.	Airfoil self-noise [12]	5	1	1503	Approximation	ASN
2.	Energy efficient [61]	8	2	768	Approximation	EE

It allows us to clearly associate the general marking of the fuzzy sets (i.e. A_i^k, B_j^k) presented in Figs. 6.7, 6.8, 6.9 and 6.10 with the linguistic labels used in the notation of rules presented in Tables 6.7 and 6.8.

The conclusions on the conducted simulations are as follows:

- The fuzzy sets for variant V2 (column a in Figs. 6.7, 6.8, 6.9 and 6.10) are characterized by poor readability in the sense of the interpretability criteria presented in Sect. 6.3. However, the systems related to these sets work with high accuracy.
- The fuzzy sets for variant V4 (column b in Figs. 6.7, 6.8, 6.9 and 6.10) have a good interpretability in the sense of the interpretability criteria presented in Sect. 6.3. The number of rules for this variant is not high (it is in the range from 3 to 4) with a good accuracy of the system (Figs. 6.11 and 6.12), which provides a good basis for interpretation of these rules.
- The fuzzy sets for variant V6 (column c in Figs. 6.7, 6.8, 6.9 and 6.10), have a very good interpretability in the sense of the interpretability criteria discussed in Sect. 6.3. In these cases the reduction of system outputs often occurs with an acceptable number of rules. Moreover, the system accuracy is also acceptable (Figs. 6.11 and 6.12).
- The results for intermediate variants V3 and V5 and extreme variants V1 and V7 are shown in Table 6.9 and in Figs. 6.11 and 6.12. They show the dependence between the system (4.28) accuracy and its interpretability. The results are (as expected) varied. This is also reflected in Figs. 6.7, 6.8, 6.9 and 6.10.
- The average number of rules 'nor', rule antecedences 'noifs', rule consequences 'noofs', inputs 'noi' and discretization points 'nod' was different for different simulation variants (Figs. 6.11 and 6.12). The values of these components decrease for cases with a better interpretability.
- The results obtained for the GFA and GIA algorithms are in all aspects better than the ones obtained for the hybrid algorithm using the genetic algorithm cooperating with the evolutionary strategy (μ, λ) (GSA) [16, 55] for $\mu = 500$ and $\lambda = 100$. The GSA algorithm was tested as a primary algorithm in order to compare the obtained results.

Table 6.3 A set of parameter values of the GFA algorithm presented in Sect. 6.2.1

Item no.	Parameter	Notation	Value
1.	Number of iterations	*Niter*	1000
2.	Number of fireworks	*Npop*	10
3.	Parameter controlling the number of generated sparks	*Nspa*	100
4.	Number of additionally generated sparks	*Nsparnd*	10
5.	Parameter limiting the minimum number of sparks	a	0.02
6.	Parameter limiting the maximum number of sparks	b	0.40
7.	Maximum amplitude of explosion	*ampMax*	0.50
8.	Minimum initial amplitude of explosion	*ampInit*	0.10
9.	Minimum final amplitude of explosion	*ampFinal*	0.01
10.	Constant preventing division by 0	ξ	0.01

Table 6.4 A set of parameter values of the GIA algorithm outlined in Sect. 6.2.2

Item no.	Parameter	Notation	Value
1.	Number of iterations	*Niter*	1000
2.	Number of countries	*Npop*	100
3.	Number of empires	*Ni*	10
4.	Random offset angle of colony	γ	0.15
5.	Probability of revolution and mutation	p_{rm}	0.15
6.	Colony importance coefficient	ζ	0.20

Table 6.5 A set of fuzzy system parameters of the form (4.28) discussed in Sect. 4.3

Item no.	Parameter description	Notation	Value
1.	Maximum number of rules	*Nmax*	7
2.	Maximum number of discretization points	*Rmax*	21
3.	Minimum value of the Dombi-norm parameters	$\underline{p}$	0.00
4.	Maximum value of the Dombi-norm parameters	$\overline{p}$	10.00
5.	Expected intersection point of fuzzy sets	c_{intB}	0.50
6.	Minimum activity of discretization points parameter	c_{intH}	0.10

6.5 Summary

In this chapter different ways of affecting interpretability of fuzzy systems designed by using evolutionary learning are discussed. Specifically, different ways of population initialization, different approaches to the evolution process and sample methods of accounting for interpretability criteria are shown. The final part of the section presents exemplary simulation results.

Table 6.6 A set of variants of the weights of the evaluation function (6.7)

Item no.	Variant	w_{ffacc}	w_{ffint}	Description
1.	V1	1.00	0.10	Focused on high accuracy
2.	V2	0.85	0.25	Focused on accuracy
3.	V3	0.70	0.40	Intermediate between V2 and V4
4.	V4	0.55	0.55	Allowing for a compromise between interpretability and accuracy
5.	V5	0.40	0.70	Intermediate between V4 and V6
6.	V6	0.25	0.85	Focused on interpretability
7.	V7	0.10	1.00	Focused on good interpretability

Table 6.7 A set of sample rules of the form (4.24) of the fuzzy system (4.28) for variant V6 and the GFA algorithm

Problem	Rule	Antecedent		Consequent	Weight
ASN	R^1: IF	(*frequency*$_{[1]}$ is *medium* |*i* AND *angle*$_{[2]}$ is *high* |*n* AND *chord length*$_{[3]}$ is *medium* |*i* AND *fs velocity*$_{[4]}$ is *low* |*v*)	THEN	(*pressure*$_{[1]}$ is *low* |*i*)	|*i*
	R^2: IF	(*frequency*$_{[1]}$ is *low* |*i* AND *angle*$_{[2]}$ is *low* |*i* AND *chord length*$_{[3]}$ is *low* |*n* AND *fs velocity*$_{[4]}$ is *high* |*i* AND *displacement*$_{[5]}$ is *low* |*i*)	THEN	(*pressure*$_{[1]}$ is *high* |*v*)	|*i*
	R^3: IF	(*frequency*$_{[1]}$ is *high* |*i* AND *chord length*$_{[3]}$ is *high* |*i* AND *displacement*$_{[4]}$ is *high* |*i*)	THEN	(*pressure*$_{[1]}$ is *medium* |*i*)	|*i*
EE	R^1:IF	(*compactness*$_{[1]}$ is *high* |*i* AND *surface area*$_{[2]}$ is *low* |*i* AND *roof area*$_{[4]}$ is *medium* |*i* AND *height*$_{[5]}$ is *near* 5.35 |*i* AND *orientation*$_{[6]}$ is *low* |*v* AND *glazing area*$_{[7]}$ is *low* |*i*)	THEN	(*heating*$_{[1]}$ is *low* |*i* AND *cooling*$_{[2]}$ is *high* |*i*)	|*i*
	R^2:IF	(*roof area*$_{[4]}$ is *low* |*i* AND *glazing area*$_{[7]}$ is *high* |*i*)	THEN	(*heating*$_{[1]}$ is *high* |*i* AND *cooling*$_{[2]}$ is *medium* |*i*)	|*i*
	R^3:IF	(*compactness*$_{[1]}$ is *low* |*i* AND *surface area*$_{[2]}$ is *high* |*i* AND *wall area*$_{[3]}$ is *near* 343.93 |*i* AND *roof area*$_{[4]}$ is *high* |*i* AND *orientation*$_{[6]}$ is *high* |*i*)	THEN	(*cooling*$_{[2]}$ is *low* |*i*)	|*i*

When designing fuzzy systems with the use of the evolutionary learning methods attention to a few important issues should be paid. First, there is a great freedom in the design of hybrid algorithms used for fuzzy system structure and parameter selection. The algorithms presented in Sects. 6.2.1 and 6.2.2 should be treated as examples. Some other sample solutions have also been presented in our previous papers [16, 17, 41, 43]. However, care should be taken so as to prevent the operation of exploration operators from compromising the operation of exploitation operators as this would result in not being able to tune the parameters of the structures indicated in the

Table 6.8 A set of sample rules of the form (4.24) of the fuzzy system (4.28) for variant V6 and the GIA algorithm

Problem	Rule	Antecedent		Consequent	
ASN	R^1:IF	($displacement_{[5]}$ is $near$ 0.01 $\|v$)	THEN	($pressure_{[1]}$ is $high$ $\|i$)	$\|i$
	R^2:IF	($frequency_{[1]}$ is $near$ 9015.01 $\|i$)	THEN	($pressure_{[1]}$ is low $\|i$)	$\|i$
	R^3:IF	($angle_{[2]}$ is $near$ 4.28 $\|i$ AND $chord\ length_{[3]}$ is $near$ 0.10 $\|i$ AND $f-svelocity_{[4]}$ is $near$ 59.09 $\|n$)	THEN	($pressure_{[1]}$ is $medium$ $\|i$)	$\|i$
EE	R^1:IF	($compactness_{[1]}$ is $near$ 0.82 $\|i$ AND $surface\ area_{[2]}$ is $near$ 717.01 $\|i$ AND $wall\ area_{[3]}$ is $high$ $\|i$ AND $height_{[5]}$ is $high$ $\|i$ AND $orientation_{[6]}$ is $near$ 4.67 $\|i$ AND $glazing\ area_{[7]}$ is low $\|i$)	THEN	($heating_{[1]}$ is low $\|i$)	$\|i$
	R^2:IF	($roof\ area_{[4]}$ is $high$ $\|i$)	THEN	($heating_{[1]}$ is $high$ $\|i$ AND $cooling_{[2]}$ is $high$ $\|i$)	$\|i$
	R^3:IF	($wall\ area_{[3]}$ is low $\|i$ AND $roof\ area_{[4]}$ is low $\|i$ AND $height_{[5]}$ is low $\|i$ AND $glazing\ area_{[7]}$ is $high$ $\|i$ AND $area\ distribution_{[8]}$ is $near$ 2.65 $\|i$)	THEN	($cooling_{[2]}$ is low $\|i$)	$\|i$

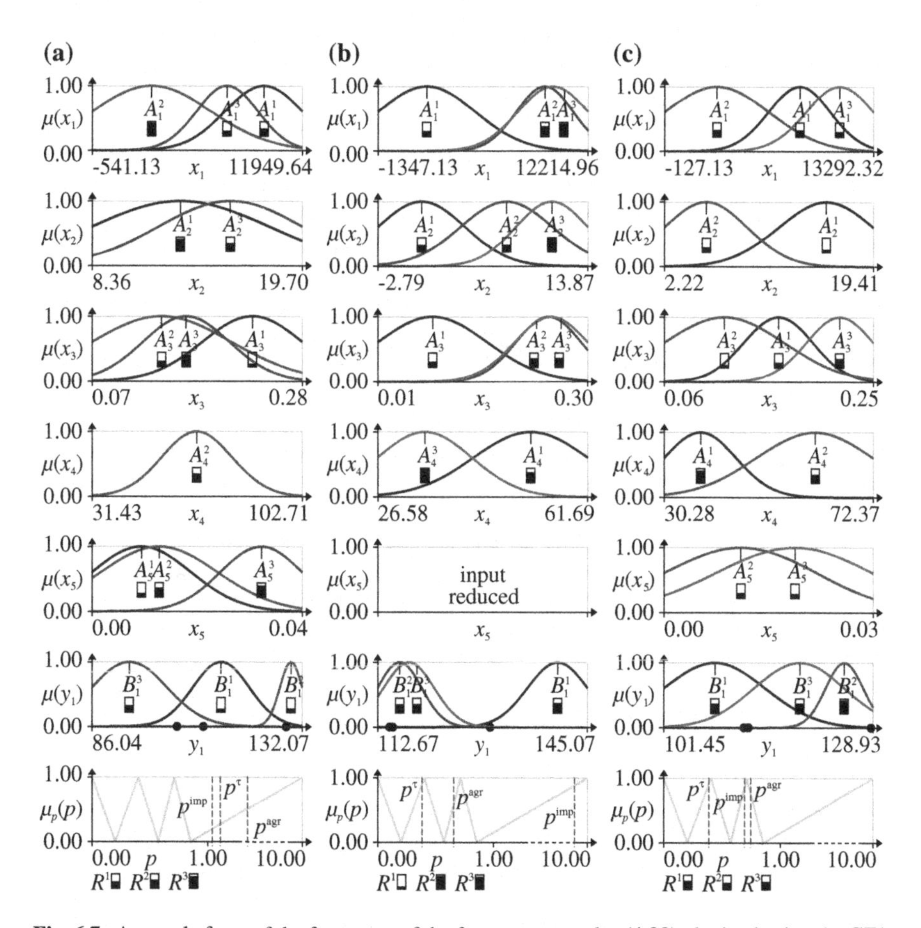

Fig. 6.7 A sample form of the fuzzy sets of the fuzzy system rules (4.28) obtained using the GFA algorithm for ASN problem and variants: **a** V2, **b** V4, **c** V6

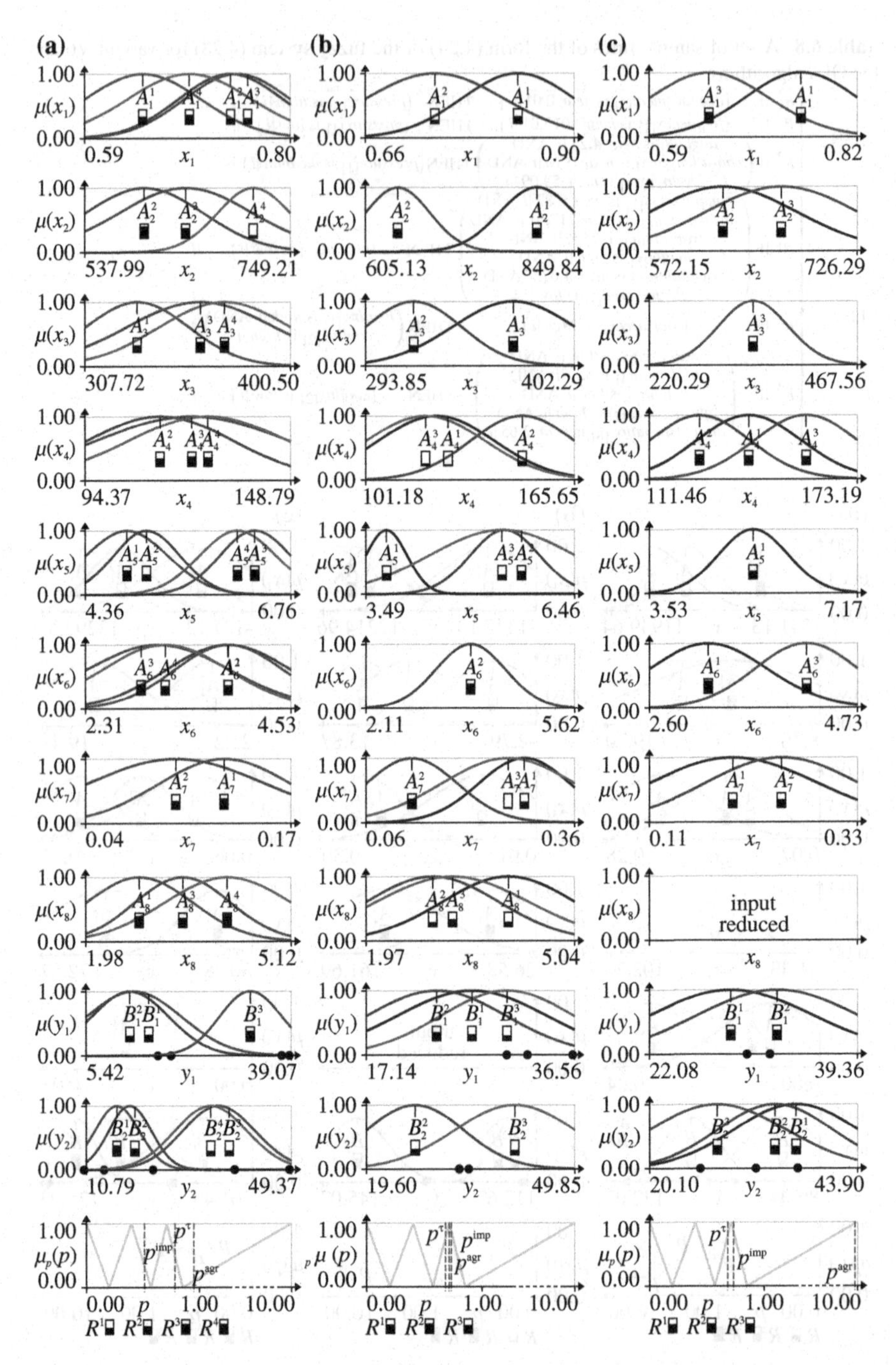

Fig. 6.8 A sample form of the fuzzy sets of the fuzzy system rules (4.28) obtained using the GFA algorithm for EE problem and variants: **a** V2, **b** V4, **c** V6

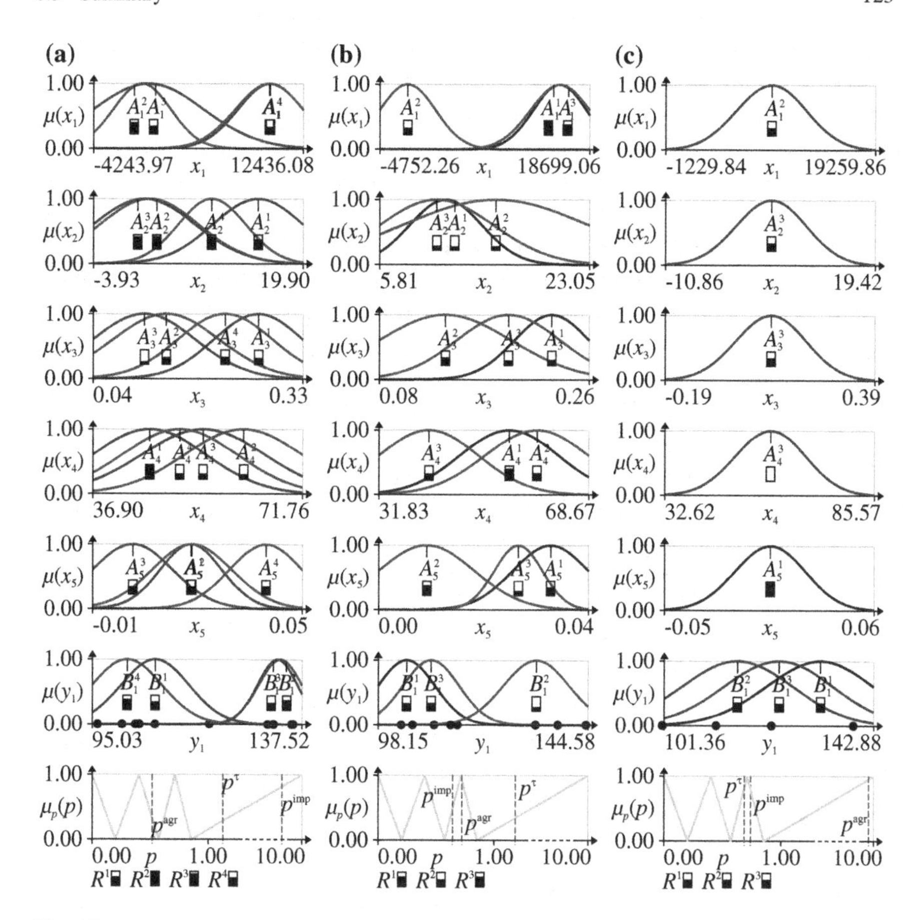

Fig. 6.9 A sample form of the fuzzy sets of the fuzzy system rules (4.28) obtained using the GIA algorithm for ANS problem and variants: **a** V2, **b** V4, **c** V6

Table 6.9 A set of values of the evaluation function (6.7) components and RMSE, averaged for 100 repetitions of the GFA, GIA and GSA algorithms. The best results for the presented variants are in bold

Problem	Algorithm	V1	V2	V3	V4	V5	V6	V7
ASN	GSA	5.6356	5.3583	8.1929	8.2468	8.3141	8.3917	8.2867
	GFA	4.7776	4.5531	6.9308	6.9039	7.0591	**6.9602**	6.9997
	GIA	**4.0560**	**4.2428**	**6.4369**	**6.2837**	**6.9364**	6.9641	**6.9371**
EE	GSA	6.5419	7.0749	7.3304	7.4335	7.7875	8.3127	8.1019
	GFA	5.6242	5.950	6.2687	6.4457	6.7250	7.0690	7.4412
	GIA	**3.9374**	**4.7172**	**5.4758**	**6.2910**	**5.8548**	**6.4350**	**6.4891**

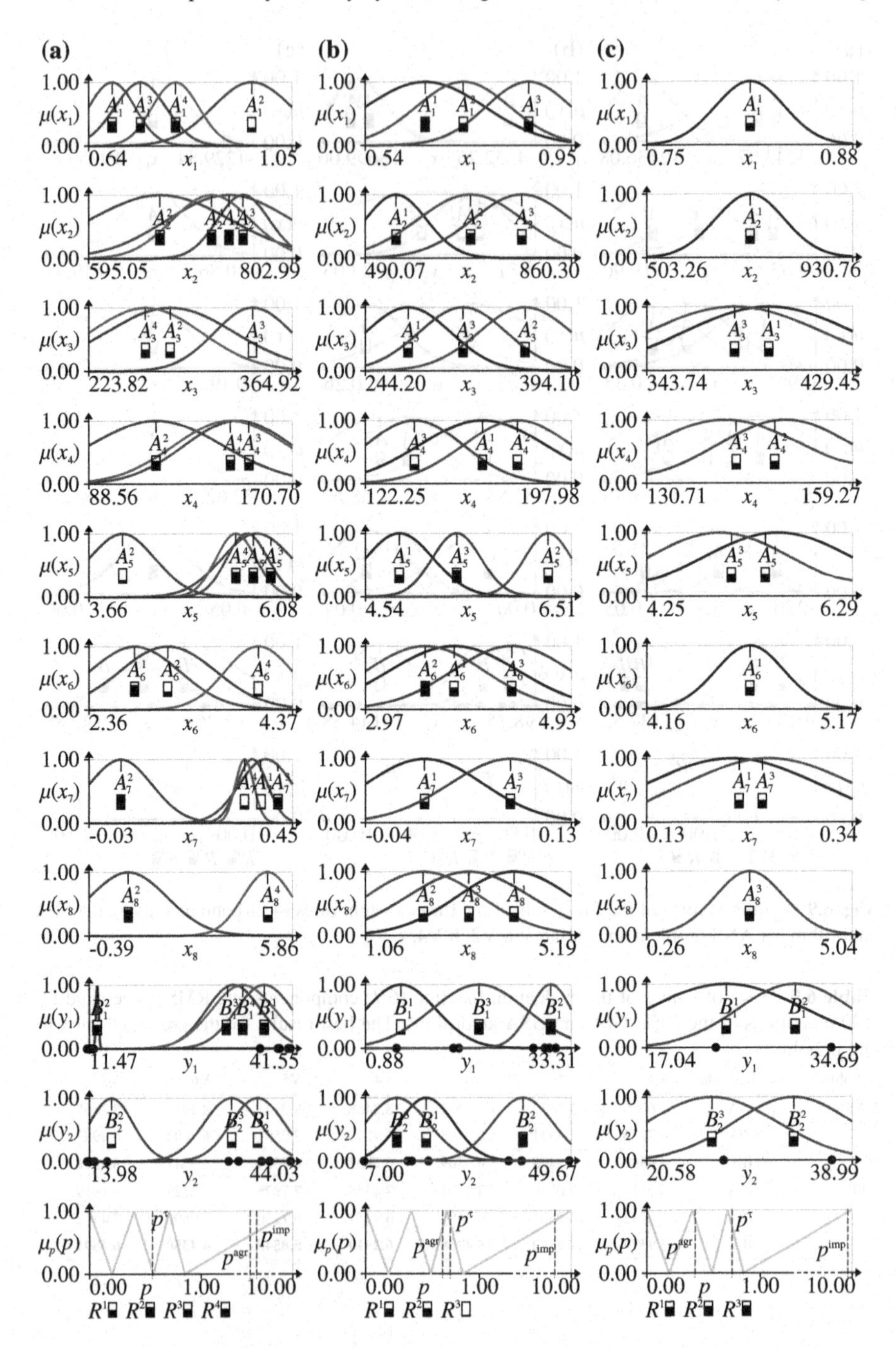

Fig. 6.10 A sample form of the fuzzy system rules (4.28) obtained using the GIA algorithm for EE problem and variants: **a** V2, **b** V4, **c** V6

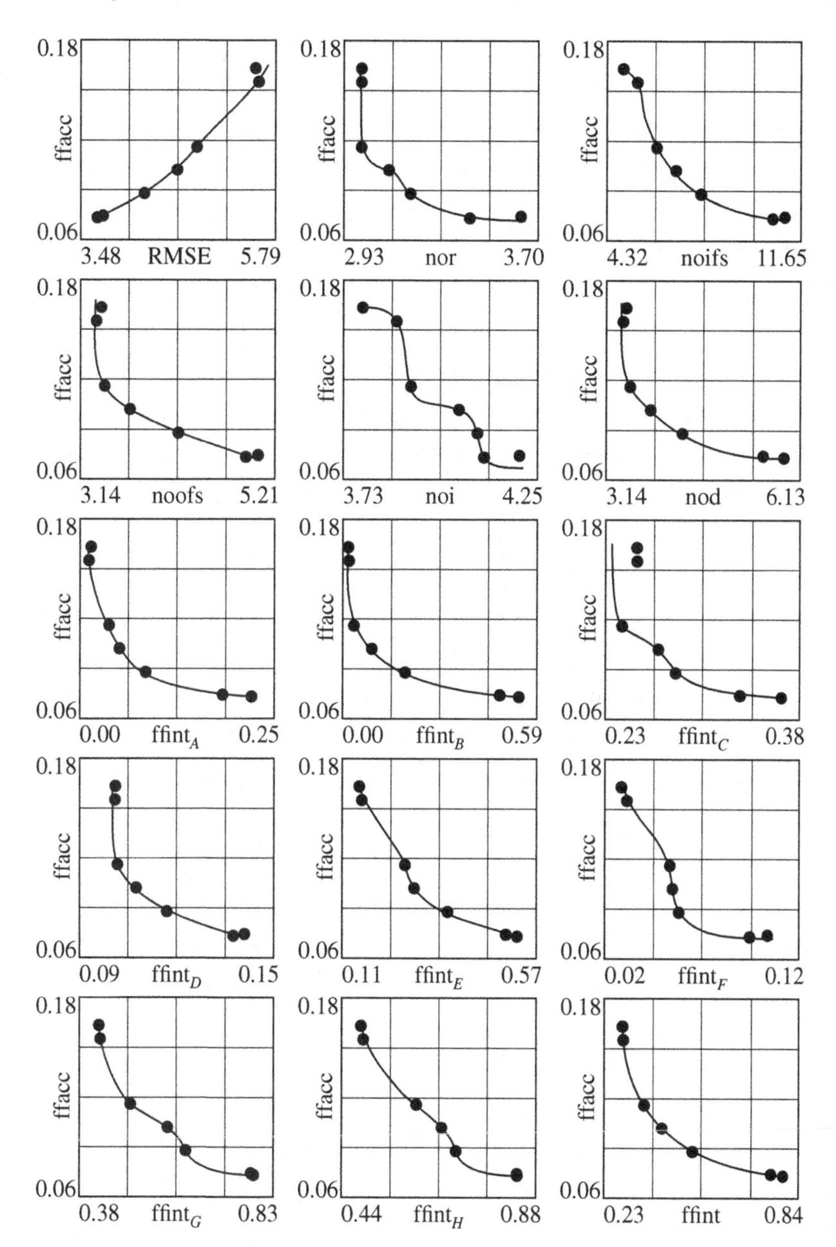

Fig. 6.11 Graphical presentation of the components of the evaluation function of the forms of (6.7) and (6.8) and the values of the functions presented in Table 6.1 averaged in the context of all the simulation problems and repetitions performed 100 times for the GFA algorithm. These values were referred to the component ffacc $(\cdot)$ regarding the accuracy of the form (2.19)

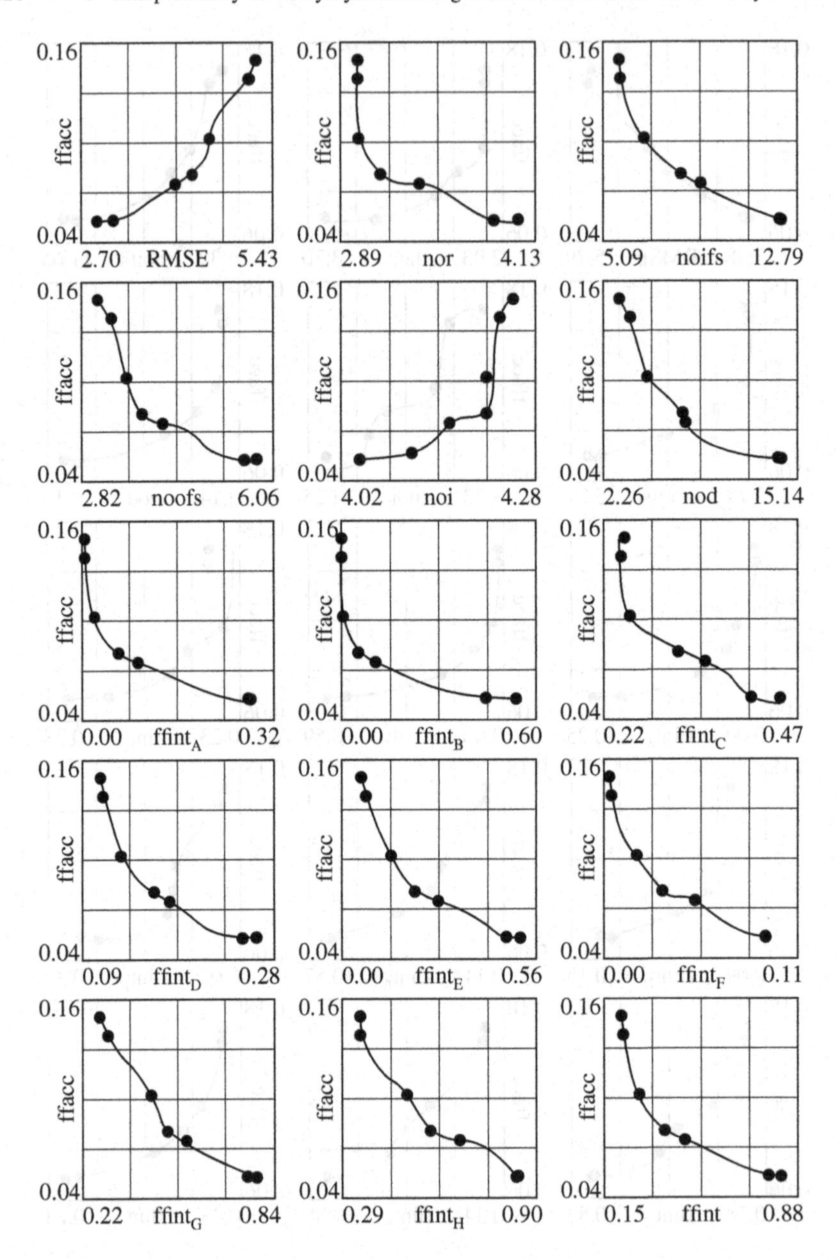

Fig. 6.12 Graphical presentation of the components of the evaluation function of the forms of (6.7) and (6.8) and the values of the functions presented in Table 6.1 averaged in the context of all the simulation problems and repetitions performed 100 times for the GIA algorithm. These values were referred to the component ffacc (·) regarding the accuracy of the form (2.19)

subsequent steps of the algorithm. The freedom in the design of hybrid algorithms also facilitates easy adaptation of the methods dedicated to processing of the Pareto fronts when needed [22]. Another consequence of the freedom in the design of hybrid algorithms is the ability to design one's own population-based algorithms modeled on the specifics of action of determined populations, sub-systems, social groups, etc.

Second, there is a great freedom of the choice of the algorithm. The most commonly used criterion for division of evolutionary algorithms takes into account the method of population processing. In this regard, among others, one-population algorithms (their sample is the genetic-firework algorithm considered in Sect. 6.2.2) and multi-population algorithms (their sample is the genetic-imperialist algorithm considered in Sect. 6.2.2) are distinguished. However, there are also other architectures of population-based algorithms, e.g. those exhibiting activity of exploitation operators.

Third, using population-based algorithms gives a large freedom in the design of evaluation function of individuals included in the population. This is reflected in the possibility of taking into consideration in this process different interpretability criteria, e.g. those considered in Sect. 6.3. The independence of the algorithm from the problem has also other advantages, e.g., it allows us to use the proposed algorithm to solve other types of problems (as shown in Chap. 9). It also allows for a flexible use of different methods of aggregating components of the evaluation function, which are characteristic for multi-criteria optimization.

Fourth, population-based algorithms do not require post-processing, which, e.g., in the case of using gradient algorithms also included the use of reduction and merging mechanisms.

It is worth noting that the solutions proposed in this section can be applied to any computational intelligence systems whose structure should be chosen individually. This eliminates the need for using the trial and error method.

Some aspects of affecting interpretability of fuzzy systems fuzzy systems designed by using evolutionary learning were partially discussed in our previous papers (see e.g. [16, 17]).

References

1. Affenzeller, M., Winkler, S., Wagner, S., Beham, A.: Genetic Algorithms and Genetic Programming: Modern Concepts and Practical Applications. Chapman and Hall/CRC, Boca Raton (2009)
2. Alba, E.: Parallel Metaheuristics: A New Class of Algorithms. Wiley-Interscience, New York (2005)
3. Alonso, J.M.: Modeling Highly Interpretable Fuzzy Systems. European Centre for Soft Computing (2010)
4. Arabas, J., Kozdrowski, S.: Population initialization in the context of a biased, problem-specific mutation. In: Proceedings of the 1998 IEEE World Congress on Computational Intelligence, The 1998 IEEE International Conference on Evolutionary Computation, pp. 769–774 (1998)
5. Arora, R.K.: Optimization: Algorithms and Applications. Chapman and Hall/CRC, Boca Raton (2015)

6. Atashpaz–Gargari, E., Lucas, C.: Imperialist competitive algorithm: an algorithm for optimization inspired by imperialistic competition. In: Proceedings of the IEEE Congress on Evolutionary Computation, vol. 7, pp. 4661–4666 (2007)
7. Back, T.: Evolutionary Algorithms in Theory and Practice: Evolution Strategies, Evolutionary Programming, Genetic Algorithms. Oxford University Press, Oxford (1996)
8. Back, T., Fogel, D.B., Michalewicz, Z. (eds.): Evolutionary Computation 1: Basic Algorithms and Operators. CRC Press, Boca Raton (2000)
9. Back, T., Fogel, D.B., Michalewicz, Z. (eds.): Evolutionary Computation 2: Basic Algorithms and Operators. CRC Press, Boca Raton (2000)
10. Banzhaf, W., Nordin, P., Keller, R.E., Francone, F.D.: Genetic Programming: An Introduction. Morgan Kaufmann, San Francisco (1997)
11. Bhuvaneswari, M.C. (ed.): Application of Evolutionary Algorithms for Multi-objective Optimization in VLSI and Embedded Systems. Springer, New York (2015)
12. Brooks, T.F., Pope, D.S., Marcolini, A.M.: Airfoil self–noise and prediction. Technical report NASA RP–1218 (1989)
13. Chambers, L.D. (ed.): The Practical Handbook of Genetic Algorithms: Complex Coding Systems. CRC Press, Boca Raton (1998)
14. Chambers, L.D. (ed.): The Practical Handbook of Genetic Algorithms: Applications. Chapman and Hall/CRC, Boca Raton (2000)
15. Chiong, R. (ed.): Nature-Inspired Algorithms for Optimisation. Springer, New York (2010)
16. Cpałka, K.: On evolutionary designing and learning of flexible neuro-fuzzy structures for nonlinear classification. Nonlinear Analysis Series A: Theory, Methods and Applications, vol. 71, pp. 1659–1672. Elsevier, New York (2009)
17. Cpałka, K., Rebrova, O., Nowicki, R., Rutkowski, L.: On design of flexible neuro-fuzzy systems for nonlinear modelling. Int. J. General Syst. **42**, 706–720 (2013)
18. Cpałka, K., Łapa, K., Przybył, A.: A new approach to design of control systems using genetic programming. Inf. Technol. Control **44**, 433–442 (2015)
19. Dasgupta, D., Michalewicz, Z. (eds.): Evolutionary Algorithms in Engineering Applications. Springer, New York (2001)
20. Dasgupta, D., Hernandez, G., Romero, A., Garrett, D., Kaushal, A., Simien, J.: On the use of informed initialization and extreme solutions sub-population in multi-objective evolutionary algorithms. In: Proceedings of the IEEE Symposium on Computational Intelligence in Multicriteria Decision-Making, pp. 58–65 (2009)
21. Deb, K., Kalyanmoy, D.: Multi-objective Optimization Using Evolutionary Algorithms. Wiley, New York (2001)
22. Deb, K., Pratap, A., Agarwal, S., Meyarivan, T.: A fast and elitist multiobjective genetic algorithm: NSGA-II. IEEE Trans. Evol. Comput. **6**, 182–197 (2002)
23. De Castro, L.N.: Fundamentals of Natural Computing: Basic Concepts, Algorithms, and Applications. Chapman and Hall/CRC, Boca Raton (2006)
24. Eiben, A.E., Smith, J.E.: Introduction to Evolutionary Computing. Springer, New York (2015)
25. Engelbrecht, A.P.: Fundamentals of Computational Swarm Intelligence. Wiley, New York (2006)
26. Falkenauer, E.: Genetic Algorithms and Grouping Problems. Wiley, New York (1998)
27. Gao, W.F., Liu, S.Y., Huang, L.L.: Particle swarm optimization with chaotic opposition-based population initialization and stochastic search technique. Commun. Nonlinear Sci. Numer. Simul. **17**, 4316–4327 (2012)
28. Gen, M., Cheng, R.: Genetic Algorithms and Engineering Optimization. Wiley-Interscience, New York (1999)
29. Goldberg, D.E.: Genetic Algorithms in Search, Optimization, and Machine Learning. Addison-Wesley Professional, Reading (1989)
30. Graham, L., Oppacher, F.: Symmetric comparator Pairs in the initialization of genetic algorithm populations for sorting networks. In: Proceedings of the 2006 IEEE International Conference on Evolutionary Computation, pp. 2845–2850 (2006)

31. Gupta, K.R.: Effect of varying the size of initial parent pool in genetic algorithm. In: Proceedings of the 2014 International Conference on Contemporary Computing and Informatics (IC3I), pp. 785–788 (2014)
32. Haupt, R.L., Haupt, S.E.: Practical Genetic Algorithms. Wiley-Interscience, New York (1998)
33. Heaton, J.: Artificial Intelligence for Humans, Volume 2: Nature–Inspired Algorithms. CreateSpace Independent Publishing Platform (2014)
34. Holland, J.H.: Adaptation in Natural and Artificial Systems: An Introductory Analysis with Applications to Biology, Control, and Artificial Intelligence. A Bradford Book (1992)
35. Ishibuchi, H., Nakashima, T., Murata, T.: Comparison of the Michigan and Pittsburgh approaches to the design of fuzzy classification systems. Electron. Commun. Jpn (Part III Fundam. Electron. Sci.) **80**, 10–19 (1997)
36. Kazimipour, B., Li, X., Qin, A.K.: A review of population initialization techniques for evolutionary algorithms. In: Proceedings of the 2014 IEEE Congress on Evolutionary Computation (CEC), pp. 2585–2592 (2014)
37. Kazimipour, B., Li, X., Qin, A.K.: Effects of population initialization on differential evolution for large scale optimization. In: Proceedings of the 2014 IEEE Congress on Evolutionary Computation (CEC), pp. 2404–2411 (2014)
38. Khan, S.S., Ahmad, A.: Cluster center initialization algorithm for K-modes clustering. Expert Syst. Appl. **40**, 7444–7456 (2013)
39. Koza, J.R.: Genetic Programming: On the Programming of Computers by Means of Natural Selection. Bradford Book (1992)
40. Krasnogor, N., Melián-Batista, B., Moreno-Pérez, J.A., Moreno-Vega, J.M., Pelta, D.A. (eds.): Nature Inspired Cooperative Strategies for Optimization (NICSO 2008). Springer, New York (2009)
41. Łapa, K., Cpałka, K.: Nonlinear pattern classification using fuzzy system and hybrid genetic-imperialist algorithm. Advances in Intelligent Systems and Computing, vol. 432, pp. 159–171. Springer, New York (2016)
42. Łapa, K., Cpałka, K.: On the application of a hybrid genetic-firework algorithm for controllers structure and parameters selection. Advances in Intelligent Systems and Computing, vol. 429, pp. 111–123. Springer, New York (2016)
43. Łapa, K., Cpałka, K., Wang, L.: New method for design of fuzzy systems for nonlinear modelling using different criteria of interpretability. Lecture Notes in Computer Science, pp. 212–227. Springer, New York (2014)
44. Łapa, K., Cpałka, K., Galushkin, A.: A new interpretability criteria for neuro-fuzzy systems for nonlinear classification. Lecture Notes in Computer Science, vol. 9119, pp. 448–468. Springer, New York (2015)
45. Łapa, K., Cpałka, K., Koprinkova-Hristova, P.: New method for fuzzy nonlinear modelling based on genetic programming. Lecture Notes in Computer Science, Springer (2016) (in press)
46. Lewis, A., Mostaghim, S., Randall, M. (eds.): Biologically-Inspired Optimisation Methods: Parallel Algorithms, Systems and Applications. Springer, New York (2009)
47. Michalewicz, Z.: Genetic Algorithms + Data Structures = Evolution Programs. Springer, New York (1996)
48. Mitchell, M.: An Introduction to Genetic Algorithms. The MIT Press, Cambridge (1996)
49. Modiri-Delshad, M., Rahim, N.A.: Fast initialization of population based methods for solving economic dispatch problems. In: Proceedings of the 3rd IET International Conference on Clean Energy and Technology (CEAT), pp. 1–5 (2014)
50. Orito, Y., Hanada, Y., Shibata, S., Yamamoto, H.: A new population initialization approach based on bordered Hessian for portfolio optimization problems. In: Proceedings of the 2013 IEEE International Conference on Systems, Man, and Cybernetics, pp. 1341–1346 (2013)
51. Pal, S.K., Wang, P.P.: Genetic Algorithms for Pattern Recognition. CRC Press, Boca Raton (1996)
52. Paul, P.V., Dhavachelvan, P., Baskaran, R.: A novel population initialization technique for genetic algorithm, circuits. In: Proceedings of the 2013 International Conference on Power and Computing Technologies (ICCPCT), pp. 1235–1238 (2013)

53. Poikolainen, I., Neri, F., Caraffini, F.: Cluster-based population initialization for differential evolution frameworks. Inf. Sci. **297**, 216–235 (2015)
54. Rahnamayan, S., Tizhoosh, H.R., Salama, M.M.A.: A novel population initialization method for accelerating evolutionary algorithms. Comput. Math. Appl. **53**, 1605–1614 (2007)
55. Rutkowski, L.: Computational Intelligence. Springer, New York (2010)
56. Sekaj, I., Perkacz, J.: Some aspects of parallel genetic algorithms with population re-initialization. In: Proceedings of the 2007 IEEE Congress on Evolutionary Computation, pp. 1333–1338 (2007)
57. Sher, G.I.: Handbook of Neuroevolution Through Erlang. Springer, New York (2013)
58. Simon, D.: Evolutionary Optimization Algorithms. Wiley, New York (2013)
59. Sivanandam, S.N., Deepa, S.N.: Introduction to Genetic Algorithms. Springer, New York (2010)
60. Tan, Y.: Fireworks Algorithm: A Novel Swarm Intelligence Optimization Method. Springer, New York (2015)
61. Tsanas, A., Xifara, A.: Accurate quantitative estimation of energy performance of residential buildings using statistical machine learning tools. Energy Build. **49**, 560–567 (2012)
62. Wang, H., Wu, Z., Wang, J., Dong, X., Yu, S., Chen, C.: A new population initialization method based on space transformation search. In: Proceedings of the 2009 Fifth International Conference on Natural Computation, pp. 332–336 (2009)
63. Yang, X.S.: Nature-Inspired Optimization Algorithms. Elsevier, Amsterdam (2014)
64. Zhang, Y., Yang, R., Zuo, J., Jing, X.: Enhancing MOEA/D with uniform population initialization, weight vector design and adjustment using uniform design. J. Syst. Eng. Electron. **26**, 1010–1022 (2015)
65. Zheng, S., Janecek, A., Tan, Y.: Enhanced fireworks algorithm. In: Proceedings of the 2013 IEEE Congress on Evolutionary Computation, pp. 2069–2077 (2013)
66. Zomaya, A.Y. (ed.): Handbook of Nature-Inspired and Innovative Computing: Integrating Classical Models with Emerging Technologies. Springer, New York (2006)

Chapter 7
Case Study: Interpretability of Fuzzy Systems Applied to Nonlinear Modelling and Control

Fuzzy systems have their limitations which result from a number of factors including a limited ability to assure their interpretability in various practical problems. If interpretability is not of paramount importance, then we can consider using methods from the "black box" group (e.g. artificial neural networks with a teacher). However, if interpretability is of a great importance, then some dedicated approaches and algorithms should be developed.

In this chapter, using methods and algorithms developed in previous chapters, two practical problems are studied: (a) problem of weakly nonlinear dynamic objects modelling (Sect. 7.1) and (b) the problem of control of a CNC machine with a jerk limit (Sect. 7.2). For both these problems exemplary simulation results (Sects. 7.1.3 and 7.2.3) and conclusions (Sects. 7.1.4 and 7.2.4) are presented. Moreover, the final part of the chapter contains its summary (Sect. 7.3) and bibliography.

7.1 Nonlinear Objects Modelling

Modelling of real objects and physical phenomena is very important from both the theoretical and practical points of view. It is used to develop control and failure detection systems, communication, analysis of chemical and biological processes, etc. [1, 4, 11, 49, 50]. Real objects are nonlinear from nature, thus the creation of their models is anything but easy. Creating a model of a linear object is much simpler. Nonlinear objects are very often modelled by means of one or several connected linear models [2, 25]. An important advantage of this approach is an easier way of creating a model which is based on the theoretical description of a known physical phenomena. A created representation of the model is interpretable, thereby these methods are treated as a white box [12, 26, 34]. However, it should be noted that because of the need to adopt simplifying assumptions these methods are often not sufficiently accurate.

One way of creating models of nonlinear objects is to observe the objects response to given input signals and an attempt to reproduce this dependence in the model

K. Cpałka, *Design of Interpretable Fuzzy Systems*, Studies in Computational Intelligence 684, DOI 10.1007/978-3-319-52881-6_7

[8, 9, 20]. Such methods are oriented primarily at achieving high accuracy during reproduction of input-output dependencies, which is, however, accomplished at the expense of the lack of interpretability of the obtained model. For this reason, this approach is referred to as a black box. In many application areas, such approach is suitable. The methods belonging to that group are, for example, neural-networks [10, 24, 33, 38, 42–44]. They are useful tools for modelling complex, nonlinear dynamic objects [26, 29, 46]. Unfortunately, in neural networks all information about an analyzed phenomenon is stored in the form of numerical weights, whose values are determined while a model is being created. As a result obtaining interpretable information about the modelled phenomenon is difficult, if not impossible at all.

As it is already mentioned in the introduction to Chap. 3, the most important methods from the point of view of interpretability are the methods from the "grey-box" group (including fuzzy systems), in which a satisfactory compromise between accuracy and interpretability can be achieved. Those models are based on physical laws describing the analyzed phenomena, while their parameters are determined by an analysis of objects behavior. Thanks to this a compromise between accuracy of a model and its interpretability can be achieved.

The approach presented in this section has been applied to weakly nonlinear dynamic objects with linear inputs and nonlinear dynamics [5]. The main features of the method can be summarized as follows:

- It is based on the linear model and generates deviations from this model. Direct use of the linear model in the areas in which the object characteristics are nonlinear can cause a sharp decline of modelling accuracy. We assume that modelling deviations from the linear model, i.e. based on linear state equations, significantly reduces or eliminates the effect of the decrease of modelling accuracy. It should be noted that the discussed method is an interesting combination of the classic approach to modelling and the approach utilizing the potential of fuzzy systems.
- It utilizes fuzzy systems to generate values of corrections to the existing linear model. Additionally, the parameters of these rules can be automatically determined by machine learning. This makes it possible to extract the information in which areas and how the linear model has been improved for greater accuracy. Machine learning can base on the use of the algorithms presented in Chap. 5 or 6.
- It can use interpretability criteria. Especially, it can take into account interpretability criteria considered in Sects. 3.4, 4.4 and 6.3.

The remainder of this chapter defines the class of weakly nonlinear objects and provides detailed assumptions concerning the outlined method of this modelling problem. Moreover, sample simulation results are presented.

7.1.1 Description of a Given Class of Objects

This section defines a class of weakly nonlinear objects and detailed assumptions concerning modelling this particular class of problems.

7.1.1.1 Modelling of Weakly Nonlinear Dynamic Objects

In a dynamic system the response depends not only on current input values but also on the values of the current state of a system. In a general case nonlinear system dynamics is described by the following equation:

$$\frac{d\mathbf{x}}{dt} = f(\mathbf{x}, \mathbf{v}), \tag{7.1}$$

where $\mathbf{x}$ is a vector of state variables, $f(\mathbf{x}, \mathbf{v})$ is a nonlinear function that represents changes of the state of an object and $\mathbf{v}$ is an input values vector. We focus on the modelling of weakly nonlinear dynamic object. Weakly nonlinear dynamic objects are those whose trend of operation is linear. Due to this, their way of operation can be approximated by linear dependencies. For such objects nonlinearities cause derogation from the linear approximation and it may results from slight changes in the parameters of certain elements of a circuit, etc. An example of a simple weakly nonlinear dynamic object is an electrical circuit consisting of real (i.e. not-ideal) elements like capacitors, resistors and inductors. In the circuit with a coil with a ferromagnetic core the inductance slightly changes in response to a change in the value of an electric current. Similarly, the resistance, inductance and capacitance change in response to temperature variations. Another example is the kinetic friction coefficient which can slightly change due to changes in the relative speed of two moving bodies. A practical example is also the asymmetry in the magnetic field distribution in electric motors, which is not included in widely used analytical models of such objects.

In the literature on the modelling of weakly nonlinear dynamic objects we can often see the following way of their approximation:

$$\frac{d\mathbf{x}}{dt} = f(\mathbf{x}, \mathbf{v}) \approx \mathbf{A} \cdot \mathbf{x} + \mathbf{B} \cdot \mathbf{v}, \tag{7.2}$$

where $\mathbf{A}$ is a system matrix (defining the system dynamics, i.e. the impact of the state variable on the state change) and $\mathbf{B}$ is an input matrix (defining the impact of the system input on the state change). The Eq. (7.2) can be applied when it is possible to determine the values of matrices $\mathbf{A}$ and $\mathbf{B}$ and the resulting accuracy is sufficient. However, since the obtained accuracy is often not sufficient, the new methods of approximation of nonlinear dynamic object are still being searched for. This is realized to simplify the analysis of the model in comparison to, for example, an analysis of the model that is based on the theoretical description of a given physical phenomenon. The simplification is the result of, among others, the possibility of using well-known methods in the fields of the control theory which have been developed for linear objects.

7.1.1.2 Modelling of Weakly Nonlinear Objects with Linear Inputs and Nonlinear Dynamics

Modelling of weakly nonlinear objects with nonlinear inputs and nonlinear dynamics can be based on the equivalent linearization technique [5]. In this method it is assumed that the general formula describing the model of the object (7.1) is expressed by the following state equation:

$$\frac{d\mathbf{x}}{dt} = \mathbf{A} \cdot \mathbf{x} + \mathbf{B} \cdot \mathbf{v} + \eta \cdot g\left(\mathbf{x}, \mathbf{v}\right), \tag{7.3}$$

where $g\left(\cdot\right)$ is a function which defines the nonlinearity of the object and η determines the impact of the function $g\left(\cdot\right)$ on the entire object. The Eq. (7.3) can be used for modelling any nonlinear object (not only weakly nonlinear objects) because the function $g\left(\cdot\right)$ theoretically can represent any nonlinearity. However, determination of the function $g\left(\cdot\right)$ for the whole range of the operation for the modelled object is difficult or not possible. For this reason the range of modelling of weakly nonlinear objects is usually limited only to the surroundings of some typical operating point $(\mathbf{x_s}, \mathbf{v_s})$. In a strictly defined range around this point the modelled object behaves in a manner similar to the linear one. Then, the influence of the component $\eta \cdot g\left(\mathbf{x}\right)$ in Eq. (7.3) is small, so the Eq. (7.3) can be simplified to the form represented by the Eq. (7.2). Such class of a system (i.e. when η is small) can be treated as weakly nonlinear system according to the explanation given in [5].

In the equivalent linearization technique the Eq. (7.3) can also be represented in an alternative form:

$$\frac{d\mathbf{x}}{dt} = \mathbf{A_{eq}} \cdot \mathbf{x} + \mathbf{B_{eq}} \cdot \mathbf{v} + e\left(\mathbf{x}, \mathbf{v}\right), \tag{7.4}$$

where matrices $\mathbf{A_{eq}}$ and $\mathbf{B_{eq}}$ describe the model of the object considered as a linear in the operating point $(\mathbf{x}_s, \mathbf{v}_s)$ and have the following form:

$$\begin{cases} \mathbf{A_{eq}} = \mathbf{A} + \mathbf{P_A} \\ \mathbf{B_{eq}} = \mathbf{B} + \mathbf{P_B}. \end{cases} \tag{7.5}$$

In the case of objects with linear inputs and nonlinear dynamics [39] the matrix $\mathbf{P_B}$ is a null matrix. The corrections matrix $\mathbf{P_A}$ is estimated for the considered operating point in such a way that the error term $e\left(\cdot\right)$ of the linear approximation is as small as possible. Finally, the model of the given weakly nonlinear dynamic object in a strictly defined range around a typical operating point $(\mathbf{x}_s, \mathbf{v}_s)$ can be written as follows:

$$\frac{d\mathbf{x}}{dt} \approx \left(\mathbf{A} + \mathbf{P_A}\right) \cdot \mathbf{x} + \mathbf{B} \cdot \mathbf{v}. \tag{7.6}$$

7.1.1.3 Modelling of Weakly Nonlinear Dynamic Objects with Linear Inputs and Nonlinear Dynamics with an Intelligent Correction of a Linear Model

The values of coefficients of the matrix $\mathbf{P_A}$ depend on the current operating point. The correction matrix values depend on the selected operating point, so they change when moving away from this point. This can significantly affect the modelling accuracy. It is the most important drawback of this kind of method of modelling.

Due to the inconvenience described earlier, in the considered method it is assumed that the values of the matrix $\mathbf{P_A}$ are not constant but they are the functions which take into account the current state $\mathbf{x}$ of the object being modelled, so $\mathbf{A_{eq}}(\mathbf{x}) = \mathbf{A}+\mathbf{P_A}(\mathbf{x})$. Due to this, these values could change with the change of the current operating point (belonging to a set of predefined operating points). Taking into account this fact, we can finally write:

$$\frac{d\mathbf{x}}{dt} \approx (\mathbf{A} + \mathbf{P_A}(\mathbf{x})) \cdot \mathbf{x} + \mathbf{B} \cdot \mathbf{v}. \tag{7.7}$$

The formula (7.7) can be treated as a base for this method of weakly nonlinear dynamic objects modelling. It was created by specifying the given problem, which allows it to be appropriately transformed in the context of fuzzy modelling taking into account interpretability aspects.

7.1.2 *Description of the Method*

The function of a fuzzy system used for modelling is generating values of the matrix $\mathbf{P_A}(\mathbf{x})$ in Eq. (7.7) (Fig. 7.1). The other features of the method can be summarized as follows:

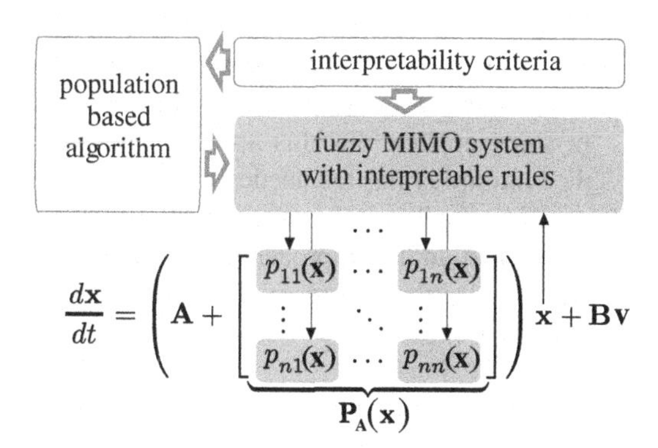

Fig. 7.1 The idea of the method for correction modelling of weakly nonlinear dynamic objects with linear inputs and nonlinear dynamics

1. It is used for the modelling of weakly nonlinear dynamic objects for which the general form of an approximated linear model is known. This means that the values of the matrices **A** and **B** are known and they result from different factors including the knowledge of the parameters of an analytical model that approximately describes the dynamics of an object. This knowledge may result from the information about physical properties of materials used for the construction of a modelled object. These properties are determined by the physical constants (such as permeability coefficient, heat capacity, etc.) and the physical characteristics (e.g. the number of turns of inductor, physical size). The knowledge of the parameters of an analytical model may also result from a previously conducted identification procedure using one of many well-known identification methods [32]. However, the problem of determining the coefficients of the matrices **A** and **B** is a separate issue and is not within the scope of this book. When the considered method has the general form of an approximated linear model, then it is able to automatically select the values of the correction matrix $\mathbf{P_A}$ in order to improve the modelling accuracy taking into account individual characteristics of a modelled real-world object. The correction matrix $\mathbf{P_A}$ might change when the current operating point changes.
2. It uses the possibilities of fuzzy systems with multiple inputs and multiple outputs. The current values of the state vector are used as the input values of the system and on this basis the system generates values of the correction matrix $\mathbf{P_A}$. The number of the system outputs depends on the dimensions of the correction matrix $\mathbf{P_A}$. This approach allows us to describe sources of nonlinearity of a given object using fuzzy rules.
3. It uses an automatic selection of values of the matrix $\mathbf{P_A}$ using the possibilities offered by the supervised learning process [35] (for example the methods outlined in Chap. 5 or 6). We assume that in order to train a system, the data from non-invasive identification of a modelled real-world object are used. It should be noted that a full description of the mathematical analysis and rigorous design methods for fuzzy control systems can be found in [16, 17].
4. It can take into account appropriately formulated interpretability criteria of fuzzy systems used for modelling the correction matrix $\mathbf{P_A}$ (also described in Sects. 3.4, 4.4 and 6.3). It is worth noting that in many approaches to nonlinear modelling fuzzy systems are used directly for modelling dependence $f(\mathbf{x}, \mathbf{v})$ in Eq. (7.1). In some applications this approach works well, but if a problem is complex, then multiple rules are needed in order to achieve a reasonable accuracy. A large number of rules makes them very difficult to analyze. As it has been already mentioned, in this approach the derogation from the approximated linear model can be described more easily with the use of fuzzy rules rather than a whole nonlinear object.

7.1.3 Simulation Results

Comments on the simulations can be summarized as follows:

- The purpose of the simulations was to illustrate possibilities of affecting fuzzy systems interpretability in weakly nonlinear dynamic objects modelling.
- The information on the used simulation problems is presented in Table 7.1. There are two sample problems from the field of nonlinear modelling, whose detailed description is presented in Tables 7.2 and 7.3.
- In the case of the first problem (HO) we assumed that the simulation time is $T = 2.0$ s and the simulation step is $dt = 1 \cdot 10^{-3}$ s (the training data set contains 2001 samples) while in the second problem (NEC) the assumption was that the simulation time is $T = 0.1$ s and the simulation step is $dt = 1 \cdot 10^{-4}$ s (the training data set contains 1001 samples). Moreover, the parameters of a nonlinear electrical circuit (Fig. 7.2) were set as follows: $R_m = 12.045\,\Omega$, $C = 500\,\mu$F, $L = 0.1$ H,

Table 7.1 A set of test problems used for the purpose of the simulation research on affecting interpretability in weakly nonlinear objects modelling problem

Item no.	Problem name	Number of input attributes	Number of output attributes	Number of sets	Problem type	Problem label
1	Harmonic oscillator with variable pulsation [27]	2	2	2001	Nonlinear modelling	HO
2	Nonlinear electrical circuit [14]	3	1	1001	Nonlinear modelling	NEC

Table 7.2 Description of the HO problem

Item no.	Name of the component describing an object	Value of the component describing an object
1	Formula describing an object	$\frac{d^2x(t)}{dt^2} + \omega^2 \cdot x(t) = 0$
2	Source of nonlinearity in an object	$\omega(x_1) = 2 \cdot \pi - \frac{\pi}{1+\lvert 2 \cdot x_1 \rvert^6}$
3	Structure of the matrix **A** in Eq. (7.7)	$\mathbf{A} = \begin{bmatrix} 0 & 2\pi \\ -2\pi & 0 \end{bmatrix}$
4	Structure of the matrix $\mathbf{P_A}$ in Eq. (7.7)	$\mathbf{P_A} = \begin{bmatrix} 0 & p_{12} \\ p_{21} & 0 \end{bmatrix}$
5	Structure of the matrix **B** in Eq. (7.7)	$\mathbf{B} = \mathbf{0}$

Table 7.3 Description of the NEC problem

Item no.	Name of the component describing an object	Value of the component describing an object
1	Formula describing an object	$\begin{cases} \frac{du(t)}{dt} = -\frac{I_s}{C} \cdot e^{a \cdot u(t)} - \frac{1}{C} \cdot i(t) + \frac{I_s + I_0}{C} \\ \frac{di(t)}{dt} = \frac{1}{L} \cdot i(t) - \frac{R_m}{L} \cdot u(t) - \frac{K_x}{L} \cdot \omega(t) \\ \frac{d\omega(t)}{dt} = \frac{K_x}{L} \cdot u(t) - \frac{K_r}{J} \cdot \omega(t), \end{cases}$
2	Source of nonlinearity in an object	$I_p = I_0 - I_s \cdot (e^{a \cdot V_p} - 1)$
3	Structure of the matrix **A** in Eq. (7.7)	$\mathbf{A} = \begin{bmatrix} -2163.86 & 2000.00 & 0.00 \\ 10.00 & -120.45 & -5.00 \\ 0.00 & 500.00 & -100.00 \end{bmatrix}$
4	Structure of the matrix $\mathbf{P_A}$ in Eq. (7.7)	$\mathbf{P_A} = \begin{bmatrix} p_{11} & 0 & 0 \\ 0 & 0 & 0 \\ 0 & 0 & 0 \end{bmatrix}$
5	Structure of the matrix **B** in Eq. (7.7)	$\mathbf{B} = \mathbf{0}$

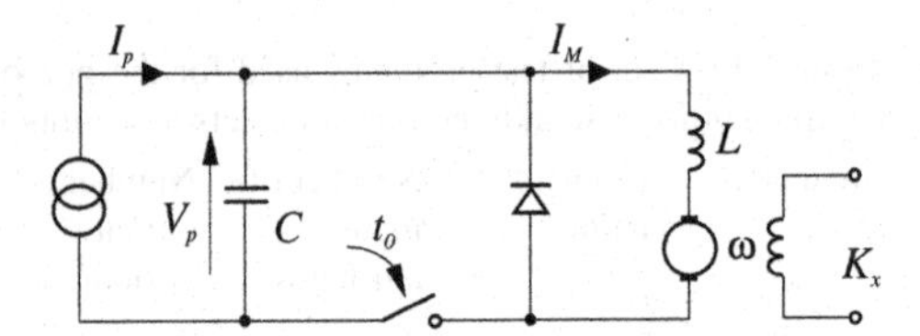

Fig. 7.2 Nonlinear electrical circuit with a solar generator and a DC drive-system diagram

Table 7.4 A set of fuzzy rules of the form (2.2) of the fuzzy system (2.12) for variant V2

HO	R^1 : IF(x_1 is *near* (+0.98)) THEN (p_{12} is *high* AND p_{21} is *low*) R^2 : IF(x_1 is *near* (−0.83)) THEN (p_{12} is *high* AND p_{21} is *low*) R^3 : IF(x_1 is *near* (+0.01)) THEN (p_{12} is *low* AND p_{21} is *high*)
NEC	R^1 : IF(u is *near* (+19.00) AND i is *near* (01.37)) THEN (p_{11} is *high*) R^2 : IF(u is *near* (+22.69) AND i is *near* (00.00)) THEN (p_{11} is *low*)

$a = 0.54\,\text{V}^{-1}$, $K_r = 0.1\,\text{Vs}^2$, $I_0 = 2\,\text{A}$, $J = 10^{-3}\,\text{Ws}^3$, $I_s = 1.28 \cdot 10^{-5}\,\text{A}$, $V_p = 22.15\,\text{V}$, $K_x = 0.5\,\text{Vs}$ [3, 14]. The values of the system matrix presented in Table 7.3 were determined using Taylor's series expansion method in point $\mathbf{x} = [22.15; 0.00; 0.00]^T$.

- In the simulations we used the Mamdani-type fuzzy system of the form (2.12) with Gaussian membership functions of the form (4.29) for input fuzzy sets and singleton membership functions of the form (2.1) for output fuzzy sets. The singleton membership functions simplify the structure of the used system because the values $\overline{y}_j$ are independent of the type of the membership function of the output fuzzy sets. Their use also makes the Mamdani-type fuzzy system equivalent to a zero-order Takagi–Sugeno type fuzzy system [13].

Table 7.5 A set of accuracies obtained for the considered test problems used in the simulations concerning affecting interpretability in weakly nonlinear objects modelling problem

Item no.	Problem label	RMSE for variant V1	RMSE for variant V2
1.	HO	0.0026	0.0090
2.	NEC	0.0026	0.0083

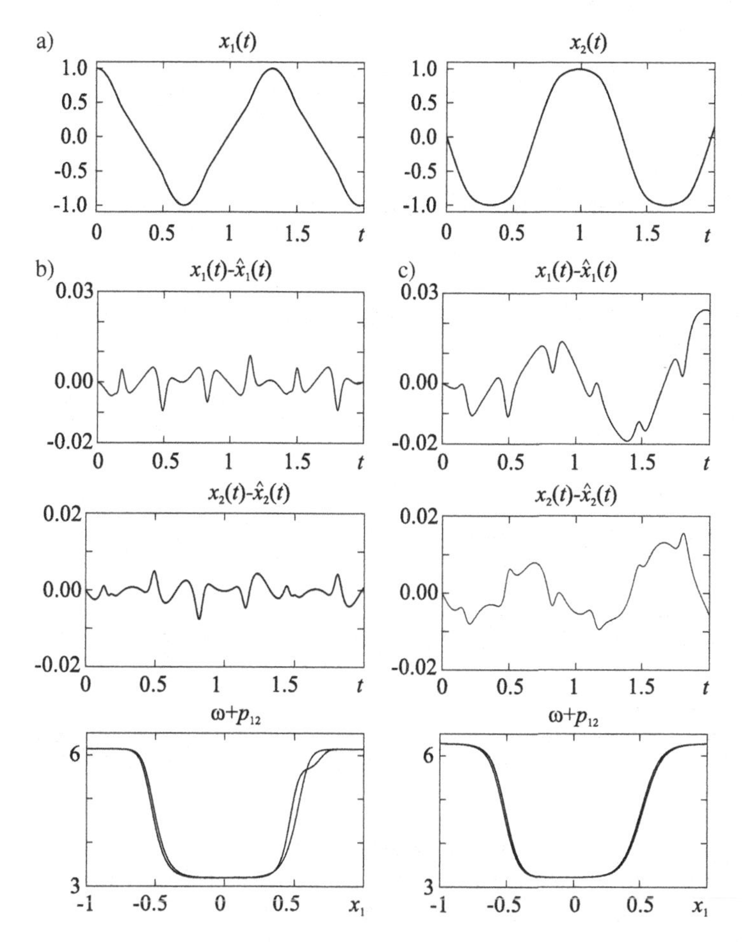

Fig. 7.3 Illustration of modelling for the HO problem: **a** reference signals, **b** signals obtained for variant V1, **c** the signals obtained for variant V2. The signals illustrate: the error obtained for signal x_1 and signal x_2, and the dependency of $\omega + p_{12}$ parameter of $\mathbf{A_{eq}}$ matrix from x_1 signal

- In the simulations we considered two variants. In the first case (V1) we focused on the accuracy of modelling, while in the second case (V2) we focused on both the accuracy of modelling and the interpretability of the fuzzy system.

The conclusions on the simulations can be summarized as follows:

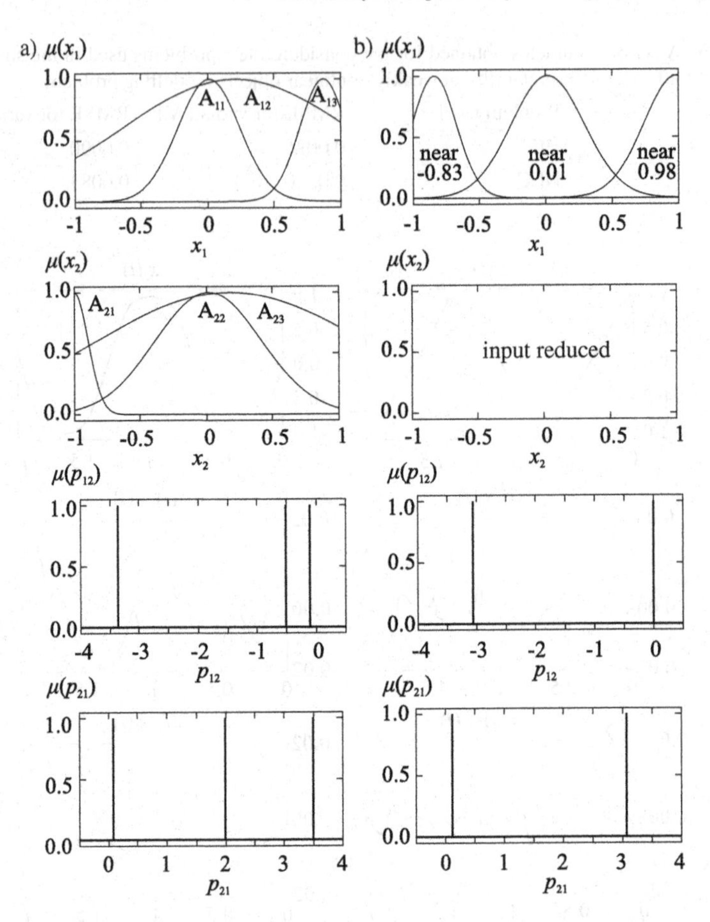

Fig. 7.4 Input and output fuzzy sets of the fuzzy system used in the HO problem in the case of: **a** high accuracy learning variant (V1) and **b** high interpretability and accuracy learning variant (V2)

- The presented method of the transformation used in weakly nonlinear objects modelling works as expected. It can be used to model dynamic weakly nonlinear objects (common in practice) with a good accuracy (Table 7.5) while effectively accounting for interpretability of the used fuzzy system (Figs. 7.4, 7.6, and Table 7.4).
- The accuracy for variant V1 aimed at the accuracy of modelling is, as expected, higher for variant V1 than for variant V2 aimed at interpretability (Table 7.5). However, the obtained modelling accuracies can be regarded as satisfactory for both considered variants (Figs. 7.3, 7.4 and 7.5).

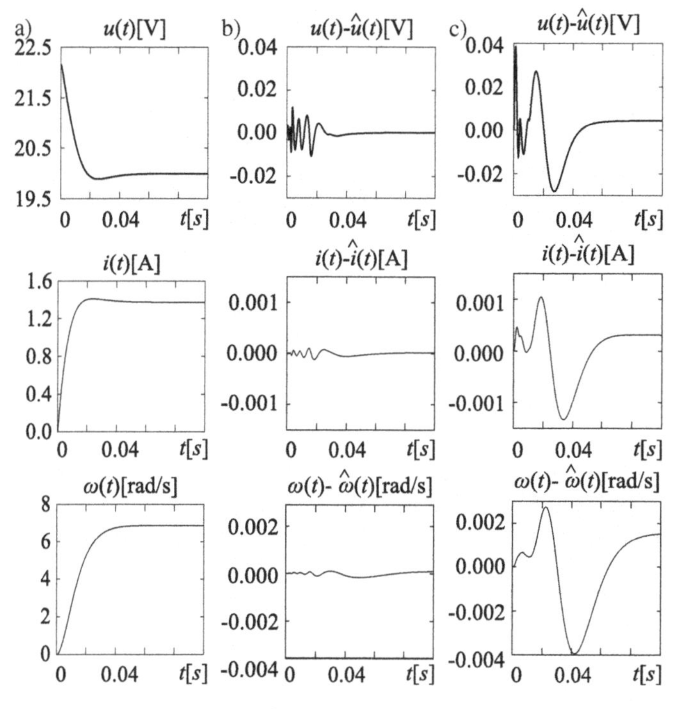

Fig. 7.5 Illustration of modelling for the NEC problem: **a** reference signals, **b** signals obtained for variant V1, **c** signals obtained for variant V2. The signals illustrate: the error obtained for signal u, the error obtained for signal i, and the error obtained for signal ω

7.1.4 Conclusions

The proposed method allows us to obtain good accuracy of weakly nonlinear objects modelling with low complexity of fuzzy systems used for modelling. Its important characteristic feature is also the fact that it is based on a hybrid approach to modelling. This makes it possible to express the model description using an equation of state variables and fuzzy rules. In this case, exclusive use of the possibilities offered by fuzzy systems seems to be ineffective in designing interpretable fuzzy systems. The method considered in this section was proposed in [3].

7.2 Control of CNC Machines with a Jerk Limit

In Computer Numerically Controlled (CNC) machines the tool high feedrate required by the High Speed Machining (HSM) technology cannot be achieved at every working point because of mechanical and electrical limitations of the machine. For example every electrical motor, when used as servo drive, has a limited output power,

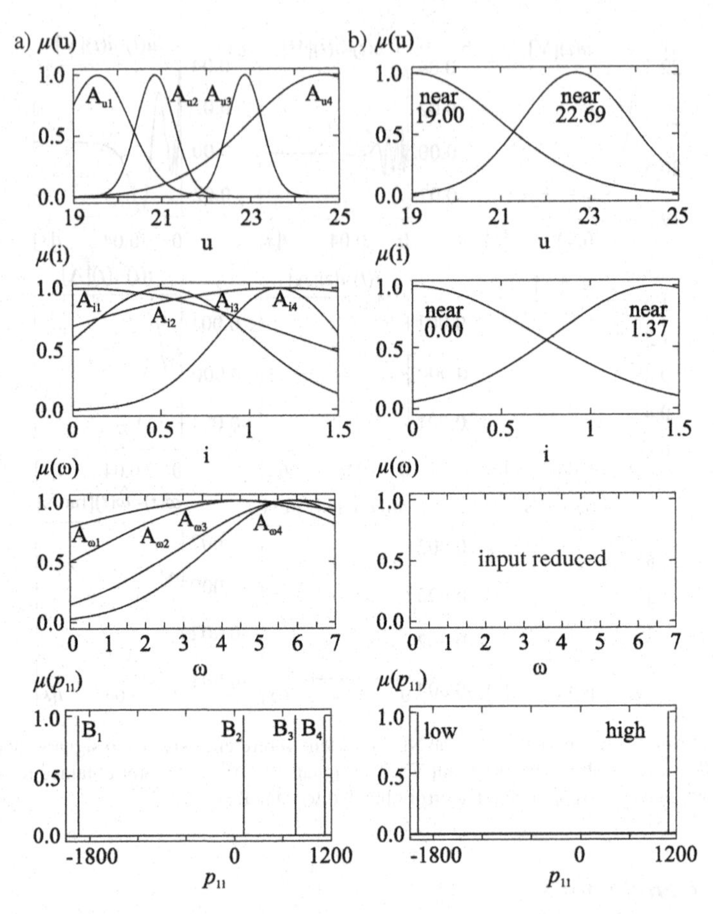

Fig. 7.6 Input and output fuzzy sets of the fuzzy system used in the NEC problem in the case of: **a** high accuracy learning variant (V1) and **b** high interpretability and accuracy learning variant (V2)

thus it can produce a limited component of centrifugal force along the toolpath. This results in a limited attainable feedrate, which depends on the current curvature of the geometrical path (Fig. 7.7) [19]. Moreover, in a CNC system, the feedrate and acceleration cannot be changed abruptly, because of the possibility of exciting the natural modes of the mechanical structure or the servo control system. A non-smooth trajectory results in premature wear of the mechanical components of a machine.

Any control system of a CNC machine should control the servo drives in such a way so as to keep the feedrate as close to the demanded value as possible, and at the same time to prevent the defined speed limits from being exceeded. Moreover, the generated trajectory should be smooth to avoid exciting the natural frequencies of the machine. A smooth trajectory can be obtained by imposing limits on the first and second time derivatives of the feedrate, which results in trapezoidal acceleration

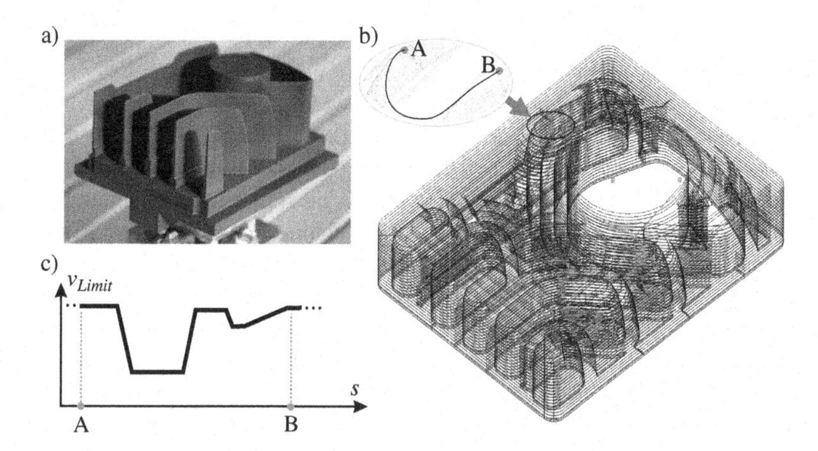

Fig. 7.7 Illustration of a sample model machined in a CNC system: **a** a graphite electrode milled on the CNC machine, **b** the geometrical path of the tool designed in the CAM (Computer Aided Manufacturing) system, **c** the velocity limit resulting from the local curvature of the path in the indicated sample fragment

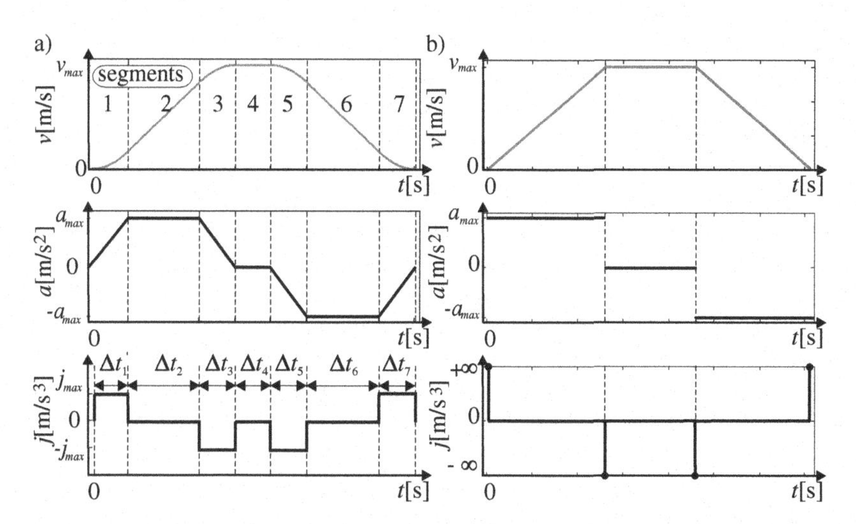

Fig. 7.8 Speed, acceleration and jerk profiles: **a** with and **b** without the jerk limit. Segments 1, 7: $jerk = j_{max}$; segments 3, 5: $jerk = -j_{max}$; segments 2, 4, 6: $jerk = 0$

profiles (Fig. 7.8a). Figure 7.8b shows a trapezoidal speed profile, which is very popular and widely used because of its simplicity. Unfortunately, it does not guarantee a high quality of machining because of discontinuities in acceleration reference values. However, if a smooth speed profile is used (Fig. 7.8a), the acceleration profile has no discontinuity and its trapezoidal form results from the jerk limit. The seven segments of that speed profile have maximal, minimal or zero values of the jerk. The trajectory presented in Fig. 7.8a describes a simple move from stop to stop, but if a more complex move is used (i.e. continuous move without stops), then these segments may occur in different sequences and/or numbers.

Online speed profile generation methods are widely investigated in literature [6, 18, 40, 41, 45]. Unfortunately, none of the reported methods were able to adjust online the generated speed profile to the changing external conditions while simultaneously limiting the value of the jerk. However, it must be emphasized that feedrate adaptation mechanisms used in high precision machines do require such feature.

We are focused on the feedrate profile generation for a set toolpath. The feedrate and its corresponding displacement are then converted to coordinates of all machine axes by using the inverse kinematics. It can be always realized even when complex interpolation methods like e.g. NURBS (Non-Uniform Rational B-Splines) based on [18, 23] are applied. Such approach enables the multi-axis interpolation task to be reduced to a one-dimensional problem. In order to simplify the terminology we will assume from now on that "speed" has the same meaning as the "feedrate" and "speed profile generation" as "trajectory generation".

The trajectory generation task is realized by an interpolator in a real time, and then calculated reference values are supplied to the servo drives. The real-time speed profile generation methods are widely investigated in literature. For example in [40] the authors discuss a real-time parametric interpolator which is able to generate continuous move along a parametric curve (e.g. linear, circular, NURBS). They take into account the machine dynamics and restrictions imposed on the feedrate along a toolpath to limit a defined error between the desired and the obtained toolpaths. Unfortunately, in this section the jerk limit is not considered at all. Similarly, in [52] the authors presented a real-time fast interpolation method which uses a look-ahead function to produce continuous move of the tool. However, the trapezoidal speed profile, presented by these authors, does not guarantee that the jerk is limited. In [47, 52] the authors extended the NURBS interpolation method taking into account the jerk limit. They proposed an iterative method for generating the trajectory. Unfortunately, their results are presented only for a relatively simple geometrical path consisting of a few splines. In practice much more complicated geometrical paths need to be used, especially in mold milling. In such case their method is rather likely to be too time consuming. In [6] there was a proposal to use a high order polynomial acceleration profile. It resulted in a smoother move but it required much more complicated numerical calculations. However, the real necessity to use such acceleration profile instead of the trapezoidal one was not proved. In [48] a method for jerk-limited trajectory planning was proposed in which the parametric interpolator is composed of a look-ahead stage and a real-time sampling stage. In that method the real-time sampling stage operates on the basis of the data calculated in the first (non real-time) stage. As a result, actually it is not the real-time algorithm because it is not able to modify the generated speed profile online. The authors in [28] focused their attention on difficulties with the online generation of polynomial-based trajectories resulting from high computational load demand- both hardware resources and processing time. They proposed hardware implementation of profile generation with a jerk limit, based on the Field Programmable Gate Array (FPGA) without using any multiplier. There was no discussion concerning a continuous move, because only rather simple moves

Table 7.6 Main features of the algorithms for jerk-limited speed profile generation

Paper	Online operation	Jerk limit	Computational complexity
Dieulot et al. [6]	No	Yes	High
Sun et al. [40]	Yes	No	Medium
Yau et al. [52]	Yes	No	Low
Yau et al. [51]	Yes	Yes	High
Tsai et al. [47]	Yes	Yes	High
Tseng et al. [48]	No	Yes	Medium
Osario-Rios et al. [28]	No	Yes	High
Lo Bianco et al. [21]	Yes	No	Low
Zheng et al. [53]	Yes	No	Low
The proposed method	Yes	Yes	Medium

from stop to stop were considered. Some authors, for example [21, 53], proposed another approach, in which the trajectory is not generated with acceleration and jerk limits, but there is a certain post-filtering method used to limit them. Such approach causes a substantial position-tracking error and it is not preferred in the reference trajectory generation used in high-precision CNC machines.

None of the methods presented in Table 7.6 were able to adjust the generated speed profile to the changing external conditions, e.g. spindle load change, in an efficient manner. As we have indicated some feedrate adaptation mechanisms used in high precision machines require such feature. Moreover, if a very complicated CAM model is being machined, it is possible that the internal memory of the interpolator does not have enough capacity to store the whole path. In such case the work must be divided into separate parts, which is unfavorable. The solution is to treat the limited memory of the interpolator as a dynamic buffer, which can be filling up while the machine is working. The incoming new data should be taken into account in the online trajectory generation method because of the necessity to generate continuous operation without unnecessary stops.

In online speed profile generation a method is developed by making use of the possibilities offered by a fuzzy system. This method is to aim at efficient generation of a smooth velocity profile for CNC machines. The unique feature of the method, which is distinguished from the other solutions, is the ability to quickly adjust the generated trajectory to changing speed limits (Fig. 7.9). In this approach it is possible to modify the demanded value of the tool feedrate while the machine is in operation. This feature is very important for operating CNC machines because of the need to protect the cutter from the brake and the spindle from the overload in high-speed machining. It is worth noting that the discussed issue has been transformed in such way that the use of a fuzzy system does not rely on replacing the solutions of trajectory generation which have proved to work well, but collaborating with them.

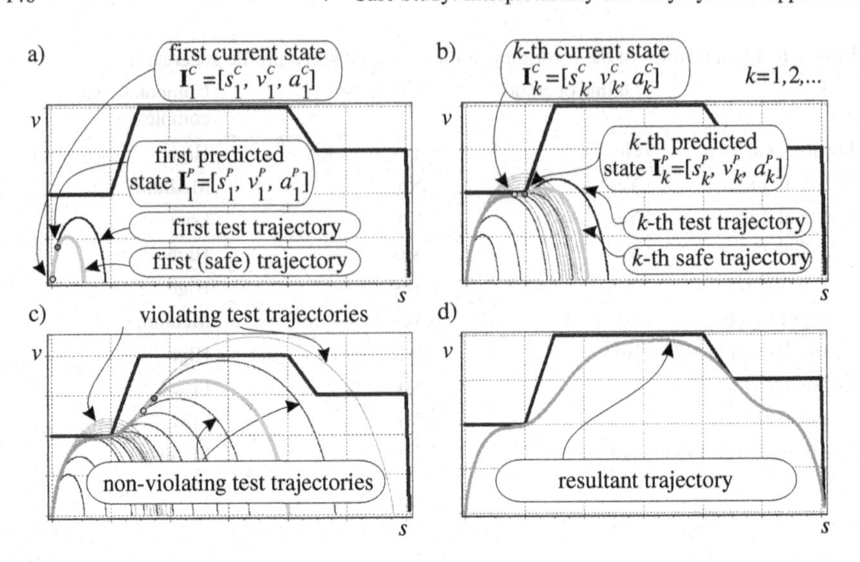

Fig. 7.9 Method for the online generation of the jerk-limited trajectory (*thick grey curve*) taking into account the feedrate limitation (the *thick black curve*). The *thin grey* and *black curves* represent test trajectories, violating and not violating velocity limitation, respectively, along the corresponding distance

7.2.1 Description of the Problem

The interpolation of a displacement, speed and acceleration as a function of time t_L with the jerk limit is based on the following well-known motion equations:

$$a(t_L) = a_0 + j_0 \cdot t_L, \tag{7.8}$$

$$v(t_L) = v_0 + a_0 \cdot t_L + j_0 \cdot \frac{t_L^2}{2}, \tag{7.9}$$

and

$$s(t_L) = s_0 + v_0 \cdot t_L + a_0 \cdot \frac{t_L^2}{2} + j_0 \cdot \frac{t_L^3}{6}, \tag{7.10}$$

where j_0 is an applied value of the jerk. Then, the interpolator state is defined by the following formula:

$$\mathbf{s} = [s, v, a]. \tag{7.11}$$

The subscript zero in (7.8)–(7.10) denotes values at a relative-time moment $t_L = 0$. Based on the Eqs. (7.8)–(7.10), an online speed profile generator is designed. The detailed flowchart of the algorithm is presented in Figs. 7.10, 7.11, 7.14, 7.15 and 7.16.

The interpolation is always based on a basis of a known current interpolator state:

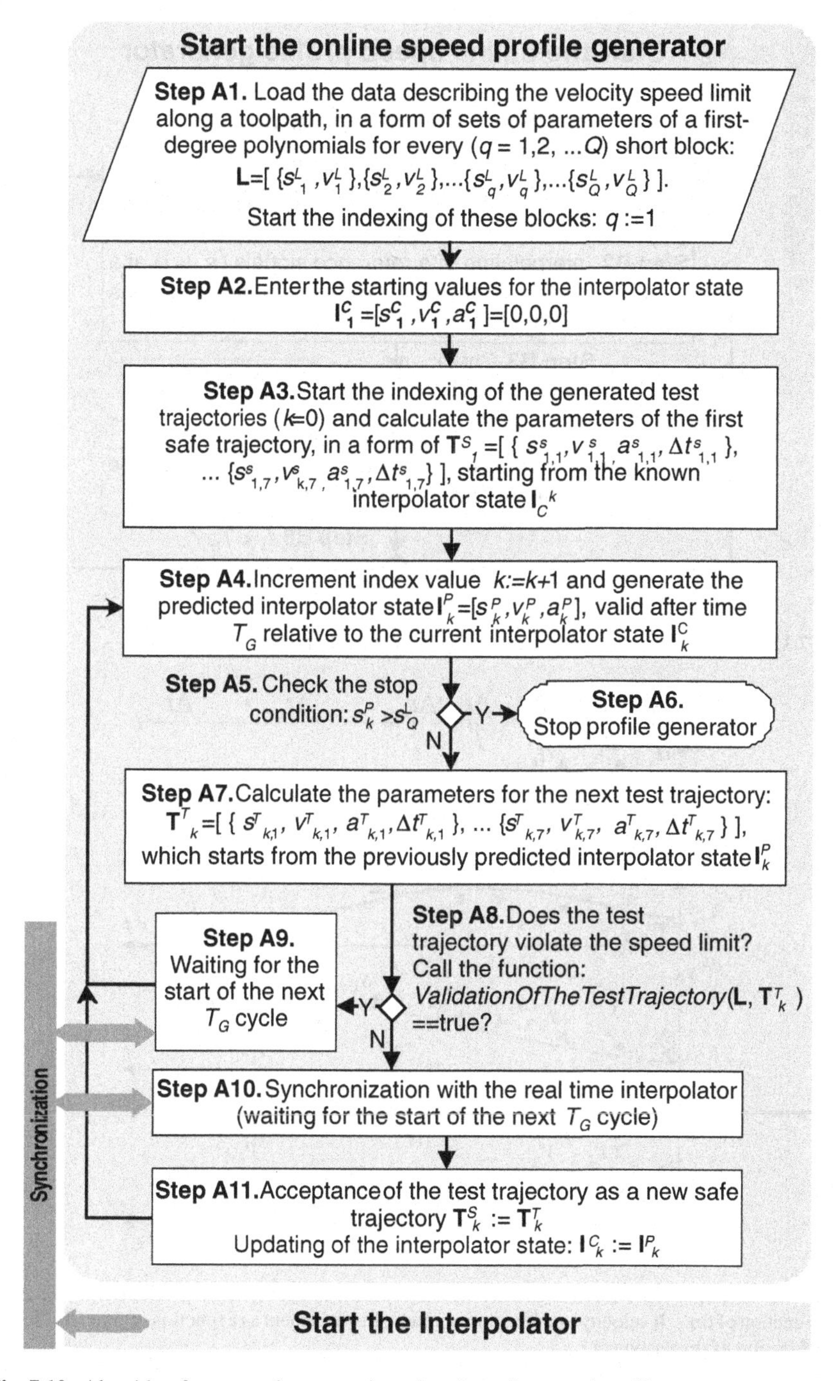

Fig. 7.10 Algorithm for generating test trajectories: the online speed profile generator

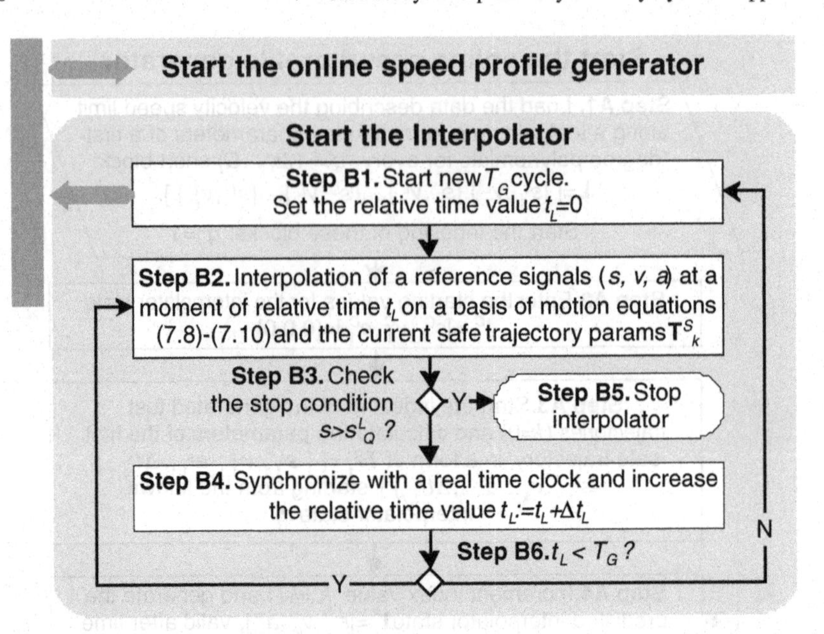

Fig. 7.11 Algorithm for generating test trajectories: the interpolator

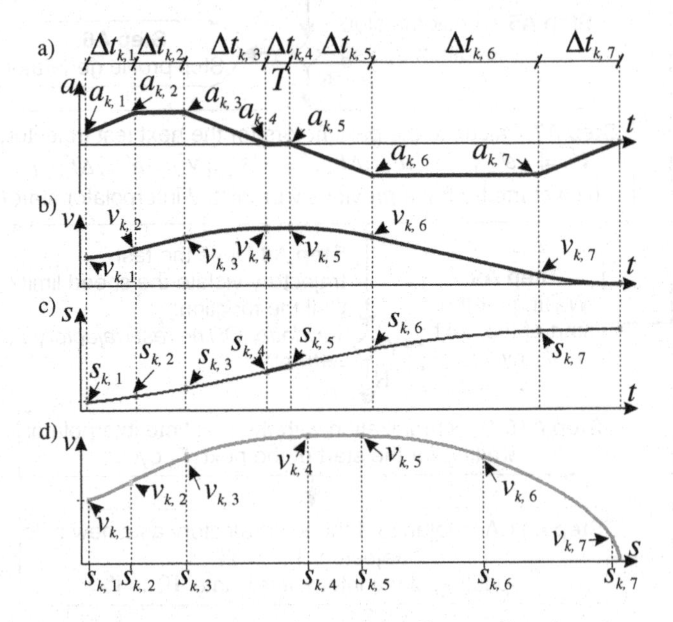

Fig. 7.12 Example of the seven-segment trajectory defined by a set of starting values: **a** acceleration as a function of time, **b** velocity as a function of time, **c** displacement as a function of time, **d** velocity as a function of displacement

$$\mathbf{I}_k^C = \left[s_k^C, v_k^C, a_k^C\right] \tag{7.12}$$

and an initially known safe trajectory (step A3 in Fig. 7.10) which is shown in Fig. 7.9a. The trajectory is defined by a set of starting values:

$$\mathbf{T}_k^S = \left[\left\{\Delta t_{k,1}^S, s_{k,1}^S, v_{k,1}^S, a_{k,1}^S\right\}, \ldots, \left\{\Delta t_{k,7}^S, s_{k,7}^S, v_{k,7}^S, a_{k,7}^S\right\}\right], \tag{7.13}$$

i.e. displacement s, velocity v, acceleration a, jerk j and the value of the time period Δt for seven successive segments as it is shown in Fig. 7.12. The safe trajectory guides the CNC machine from the starting point to the stop guaranteeing the velocity, acceleration and jerk limits.

We can easily predict (step A4 in Fig. 7.10) a future state of the interpolator (in a time distance T_G from the current moment) along the safe trajectory:

$$\mathbf{I}_k^P = \left[s_k^P, v_k^P, a_k^P\right]. \tag{7.14}$$

Treating this predicted state as a starting point we can generate one test trajectory (step A7 in Fig. 7.10) defined by the following set of parameters:

$$\mathbf{T}_k^T = \left[\left\{\Delta t_{k,1}^T, s_{k,1}^T, v_{k,1}^T, a_{k,1}^T\right\}, \ldots, \left\{\Delta t_{k,7}^T, s_{k,7}^T, v_{k,7}^T, a_{k,7}^T\right\}\right]. \tag{7.15}$$

The test trajectory has the task to speed up the move a little when compared with the move resulting from the safe trajectory (Fig. 7.13). This speed-up can be easily done by applying a non-zero time period (Δt_1^T, Δt_2^T, Δt_3^T) values in a seven-segment speed profile. In our case the sum of these three parameters was chosen experimentally and is equal to the value of parameter T_G. After this time interval the move should be immediately slowing down to a stop in the way defined by the

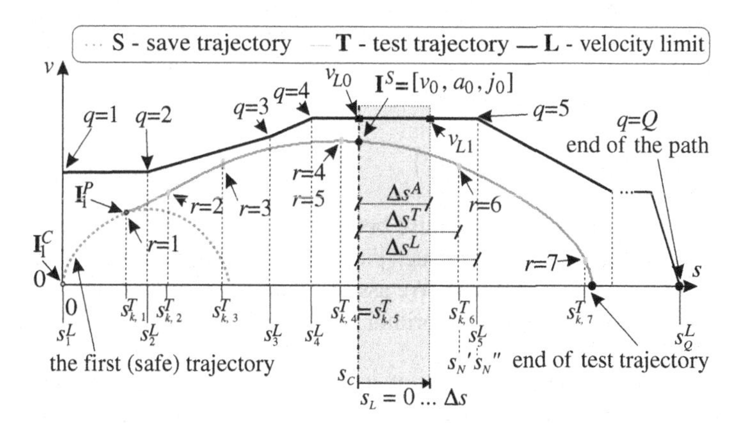

Fig. 7.13 Example of a test trajectory as a function of distance and intersections of its segments with speed constraints blocks. The *grey* area shows the currently analyzed sectors in the iterative validation algorithm of the test trajectory

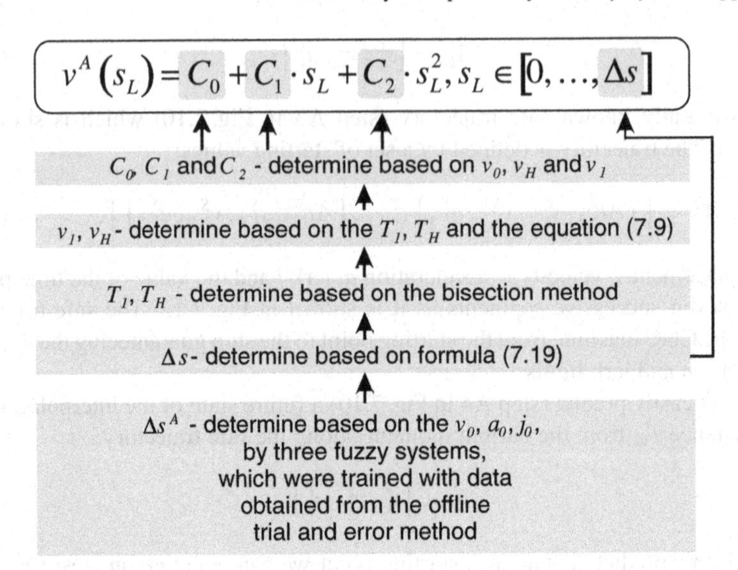

$v^A(s_L) = C_0 + C_1 \cdot s_L + C_2 \cdot s_L^2, s_L \in [0, \ldots, \Delta s]$
C_0, C_1 and C_2 - determine based on v_0, v_H and v_1
v_1, v_H - determine based on the T_1, T_H and the equation (7.9)
T_1, T_H - determine based on the bisection method
Δs - determine based on formula (7.19)
Δs^A - determine based on the v_0, a_0, j_0,
by three fuzzy systems,
which were trained with data
obtained from the offline
trial and error method

Fig. 7.14 Flowchart illustrating how the fuzzy system efficiently aids the quadratic approximation of the test trajectory as a function of the displacement in the subsequent segments

seven-segment trajectory using the segments marked as 5, 6 and 7 (Fig. 7.8a). The calculations required to determine the parameters of the seven-segment trajectory are extensively presented in the literature [7, 15, 19, 22] and will not be presented in this book.

The generated test trajectory has to be validated (step A8 in Figs. 7.10, 7.15 and 7.16) so as to determine whether it violates or not the velocity limit along the toolpath (the thick black curve in Fig. 7.9). The speed limit depends on the local curvature of the geometrical path designed by a CAM system [19] and this dependency can be approximated by any piecewise function, for example, by the zero or higher-order polynomial [15]. We use the first-order polynomial described by the following sets of reference knots (Figs. 7.7c and 7.13):

$$\mathbf{L} = \left[\{s_1^L, v_1^L\}, \{s_2^L, v_2^L\}, \ldots, \{s_q^L, v_q^L\}, \ldots, \{s_Q^L, v_Q^L\} \right], \tag{7.16}$$

which gives a satisfactory compromise between accuracy and computational complexity. The number of blocks Q of such piecewise curve depends on the complexity and length of the geometrical path. We assume that the piecewise curve was determined in advance by a separate algorithm [51], which will not be discussed in this book.

If a validation algorithm determines that the test trajectory violates the speed limit, this trajectory will be discarded (the thin grey curve in Fig. 7.9) which is shown in the block diagram as step A9. Otherwise, this trajectory will be a new safe trajectory valid after T_G time period (steps A10 and A11 in Fig. 7.10). This procedure is repeated in successive time periods and consecutive test trajectories are generated, each starting

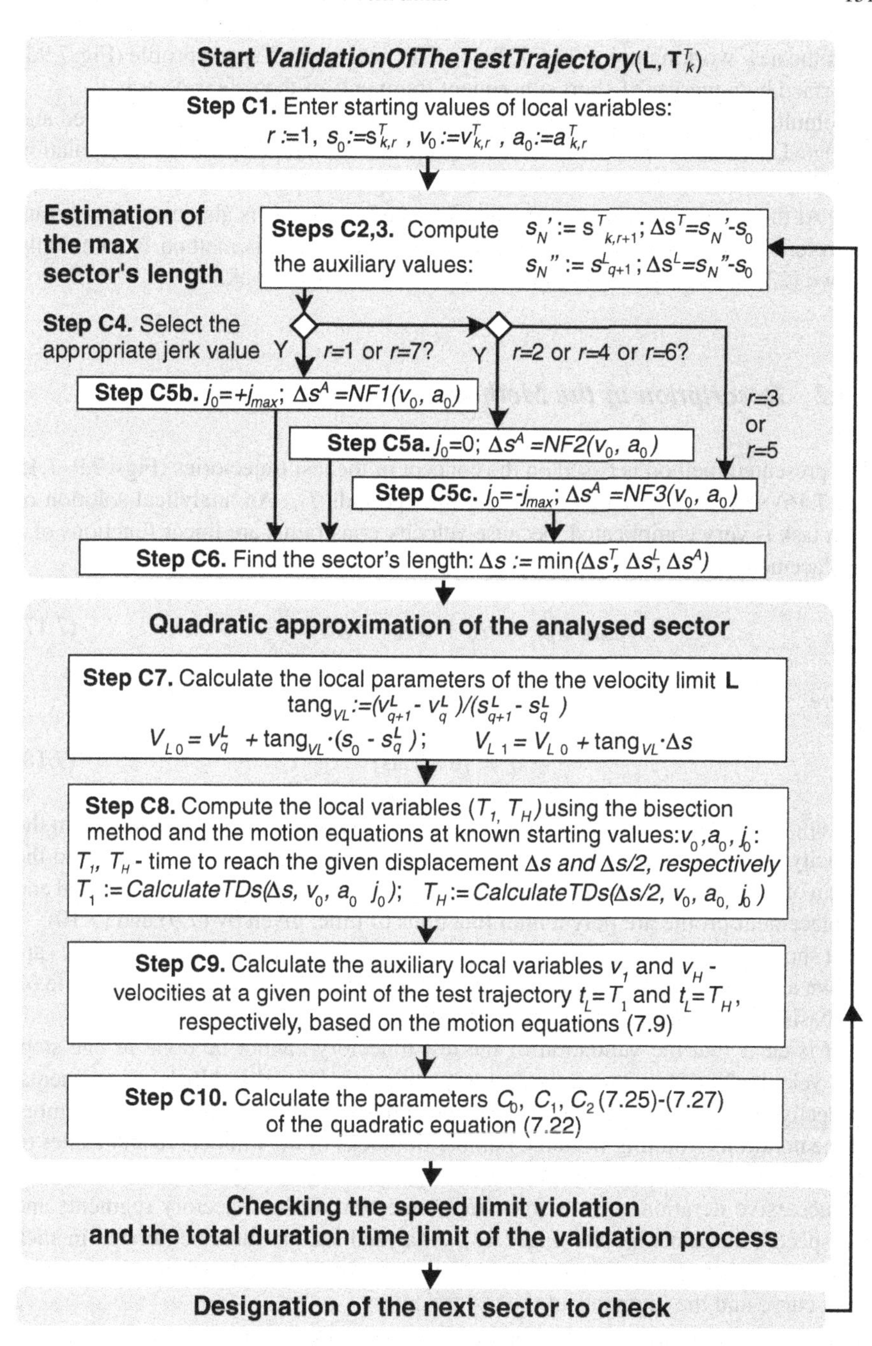

Fig. 7.15 Flowchart illustrating validation of the test trajectory: estimation of the maximal sector's length and quadratic approximation of the analyzed sector

from the new working point (Fig. 7.9b, c). The final smooth speed profile (Fig. 7.9d) is formed by a merger of short subsequent fragments of the safe trajectory.

Simultaneously, in a real time, when the test trajectories are being generated and validated, a motion controller is working (steps B1–B6). It performs the interpolation of a displacement, speed and acceleration along a toolpath in the fixed-time steps Δt_L. At this point the proper (adequate) inverse kinematics is also used to generate the reference values for all the machine's servo drives. This method is commonly known [22, 30] and hence, it will not be presented in this book.

7.2.2 Description of the Method

The presented method is based on the concept of the test trajectories (Figs. 7.9, 7.15 and 7.16) which are generated in fixed-time periods T_G. An analytical solution of such task is very complicated, because velocity constraints are linear functions of a displacement given by:

$$v_{Limit}(s_L) = v_{L0} + (v_{L1} - v_{L0}) \cdot \frac{s_L}{\Delta s}, \tag{7.17}$$

where

$$s_L = [0 \ldots \Delta s], \tag{7.18}$$

and v_{L0}, v_{L1} are the parameters of a currently analyzed sector Δs resulting from the velocity constraint curve (Fig. 7.13), t_L and s_L are time and position relative to the origin of the considered sector (the grey area in Fig. 7.13), while generated speed and displacement profile are polynomial functions of time, given by (7.9) and (7.10).

It should be noted that in Fig. 7.12d as well as in Fig. 7.13 the velocity profiles are shown as a function of a displacement. Creating these figures was only possible on the basis of the performed iterative simulation.

It is clear that the validation of the test trajectory cannot be done in one step. The velocity limit curve and the test trajectory are defined in blocks or segments, respectively. As a result the validation algorithm must be an iterative, with the number of the iterations resulting from the number of blocks of the limit curve and values of the parameters of the test trajectory. The length of the currently analyzed sector (Δs) in successive iterations results from an intersection of the trajectory segments and the speed constraints blocks (Fig. 7.13). Analyzed sectors must be iterated in such way so that they do not cross the boundaries marked by the blocks of the velocity limit curve and the segments of the test trajectory.

The presented method, which is based on a quadratic approximation of the test trajectory, needs to satisfy the following limitation: the value of Δs cannot be greater than the parameter Δs^A, which is explained further on in this section, otherwise the approximation accuracy will be poor. As a result, the sector's length is determined

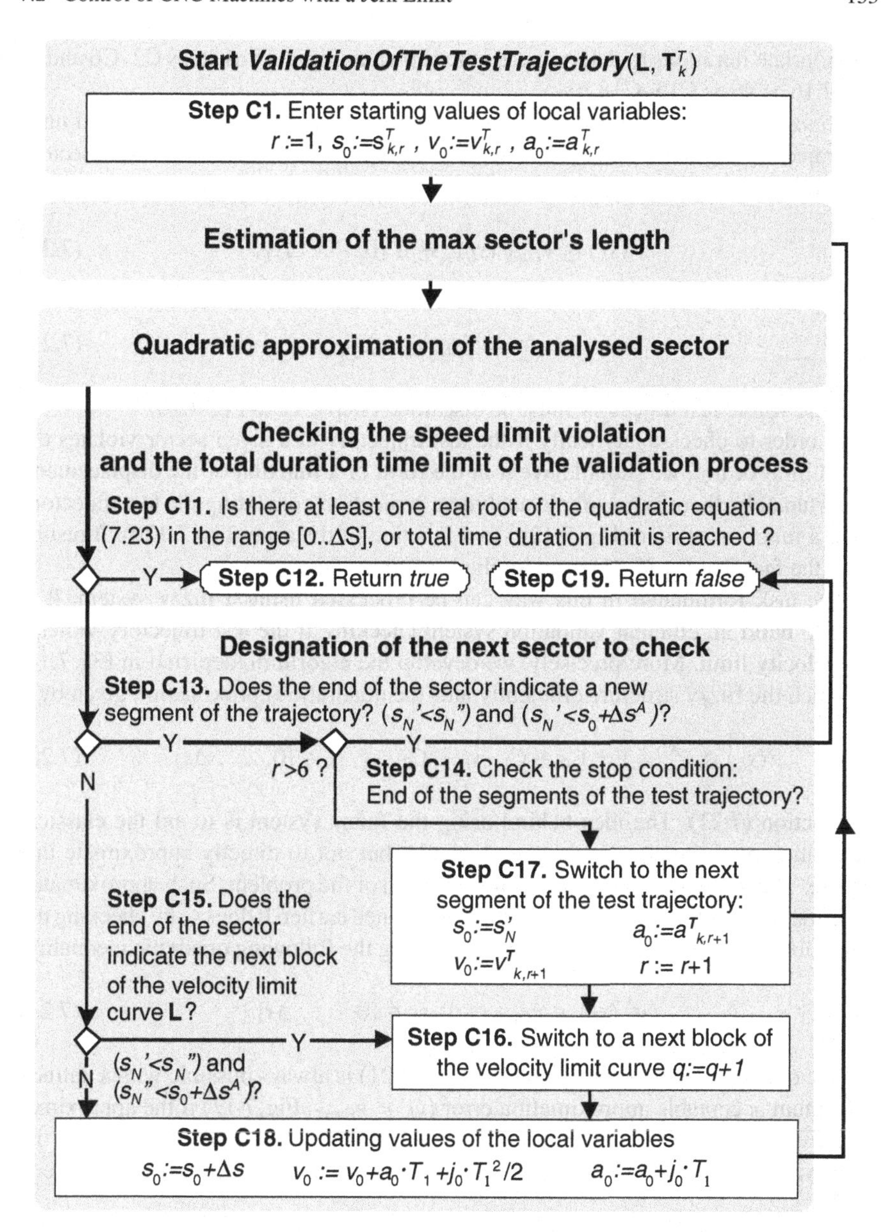

Fig. 7.16 Flowchart illustrating validation of the test trajectory: checking the speed limit violation and the total duration time limit of the validation process and the designation of the next sector to check

as the minimal value of three values (Δs^A, Δs^T, Δs^L), as it is shown in Fig. 7.13, i.e.:

$$\Delta s = \min\left(\Delta s^A, \Delta s^T, \Delta s^L\right). \tag{7.19}$$

Appropriate iterations to determine Δs are shown in Fig. 7.15 as steps C2–C6 and in Fig. 7.16 as steps C13–C18.

To validate the whole test trajectory, all successive sectors must be tested until the trajectory segments (7.15) end (step C19 in Fig. 7.16). If any of the tested sectors violates the speed limit:

$$v(s_L) \leq v_{Limit}(s_L), s_L \in [0, \ldots, \Delta s], \tag{7.20}$$

where

$$v(s_L) = v(s_L, v_0, a_0, j_0), \tag{7.21}$$

then the whole test trajectory must be discarded (step C12 in Fig. 7.16)

In order to check analytically if the test trajectory at a given sector violates the speed limit or not, we should have it in the form of a function of the displacement. Unfortunately, there is no simple analytical projection converting the test trajectory from a function of time (Fig. 7.12b) to a function of distance (Fig. 7.12d). It results from the fact that Eq. (7.21) is an implicit function.

The task formulated in this way can be processed using a fuzzy system. It is used to build an efficient validation system checking if the test trajectory violates the velocity limit. More precisely, we develop the algorithm, depicted in Fig. 7.14, in which the fuzzy structure efficiently aids the quadratic approximation given by:

$$v(s_L) \approx v^A(s_L) = C_0 + C_1 \cdot s_L + C_2 \cdot s_L^2, s_L \in [0, \ldots, \Delta s] \tag{7.22}$$

of function (7.21). The idea behind using the fuzzy system is to aid the classical quadratic approximation of the speed profile, but not to directly approximate this profile. Therein lies the proposed transformation of the problem. Such approximated function reduces the complex calculation mentioned earlier. It does so by checking the condition (7.20) through a simple task of solving the following quadratic inequality:

$$v^A(s_L) \leq v_{Limit}(s_L), s_L \in [0, \ldots, \Delta s]. \tag{7.23}$$

The quadratic approximation of function (7.21) is always possible with a defined maximum acceptable approximation error ($v_E < v_{Emax}$, Fig. 7.17) if the approximation distance Δs does not exceed Δs^A, which depends on the current curvature of the approximated function (7.21). The value of Δs^A depends on the three parameters, i.e.:

$$\Delta s^A = \Delta s^A(v_0, a_0, j_0), \tag{7.24}$$

which fully defines the interpolator state at the origin of the analyzed sector (Fig. 7.13).

Unfortunately, this dependency is not known in advance and can only be obtained by the trial and error method based on many repeated iterative simulations using motion equations (7.8)–(7.10). Since the trial and error method is very time consuming, it is not suitable for use in the validation system. Fortunately, we can use the fuzzy structure to approximate dependency (7.24) in an efficient manner.

Finally, if we know the value of Δs^A, we can approximate function (7.21) in the form of (7.22) making additional analytical calculations resulting from the use of Dirichlet boundary conditions, i.e.:

$$C_0 = v_0, \tag{7.25}$$

$$C_1 = -\frac{3 \cdot v_0 + v_1 - 4 \cdot v_H}{\Delta s}, \tag{7.26}$$

and

$$C_2 = \frac{2 \cdot v_0 + 2 \cdot v_1 - 4 \cdot v_H}{\Delta s^2}. \tag{7.27}$$

The presented boundary conditions require that the values of the approximated function (7.21) and their quadratic approximation (7.22) be equal at the start point (v_0), half point (v_H) and end point (v_1) of the approximation distance Δs (Fig. 7.17). The velocity v_0 at the origin of the sector is already known (Fig. 7.13), but the values of v_H and v_1 are not known and should be determined here. At first, the corresponding values of the relative times (ΔT_H) and (ΔT_1) must be determined. It can be easily done by using a commonly known bisection method with the utilization of the motion equation (7.10) and the values of the interpolator state described by:

$$\mathbf{I}^S = [s_0, v_0, a_0] \tag{7.28}$$

at the origin of the analyzed sector. The move defined by the test trajectory is progressive (i.e. $v(t_L) \geq 0$) within the considered range, so the bisection method with over a dozen simple iterations is sufficient to obtain a satisfactory accuracy (step C8 in Fig. 7.15). In the bisection method the maximum value of ΔT_{max} for the search algorithm is set to a minimal positive value of the relative time at which the velocity described by formula (7.9) reaches the value equal to zero. If the velocity does not reach zero for any positive value of the relative time, then the value of ΔT_{max} is set to a reasonable limit equal to one second. The minimal value for the search algorithm is set to zero.

If the values of (ΔT_H) and (ΔT_1) are calculated, then the values of v_H and v_1 can be easily determined (step C9) on the basis of the motion equation (7.9). Finally, at step

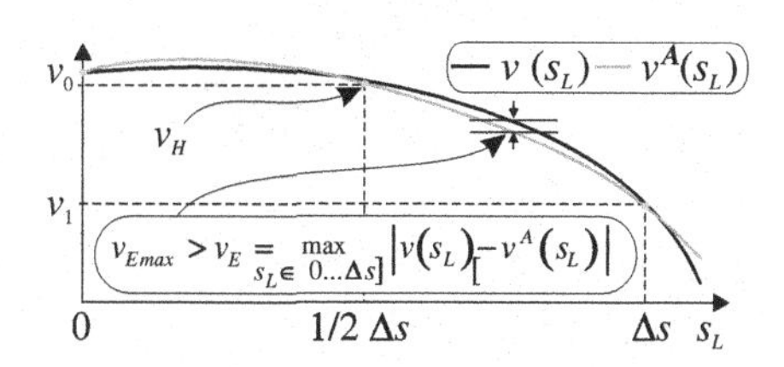

Fig. 7.17 Quadratic approximation of a velocity profile as a function of displacement

C10 of the presented algorithm, parameters C_0, C_1 and C_2 of the quadratic function (7.22) can be simply calculated using formulas (7.25)–(7.27). As a result, we can use the quadratic inequality (7.23) instead of the complicated formula (7.20), which significantly simplifies the trajectory validation algorithm. In the next step (C11) we need to solve the quadratic inequality (7.23) to check whether the analyzed sector of the test trajectory violates the speed limit or not. It can be easily done by checking if the adequate quadratic equation has at least one real root in the range $[0, \Delta_s]$. Since v_0 always takes the value smaller than v_{L0} (Fig. 7.13), the real root within the range $[0, \Delta_s]$ is the velocity violation point. As it was previously explained, in such case validation of the test trajectory is completed (step C12) with the result equal to *true*. This procedure is also terminated with the result equal to *true* if the total duration time limit (T_G) of the validation procedure is reached. In another case the steps C13–C19 are performed to switch to the next sector in the iterative validation algorithm.

7.2.3 Simulation Results

The comments on the simulations can be summarized as follows:

- The purpose of the simulations was to show possibilities of affecting interpretability of fuzzy systems in the issue of control with a jerk limit. The idea of the developed method that the fuzzy system can effectively support the validation of a trajectory without e.g. having to generate it.
- In the validation process three different independent fuzzy systems of the form (4.28) were used for approximation of the highly nonlinear dependency (7.24). In that case each of the three fuzzy systems (FS1, FS2 and FS3) approximate a highly nonlinear function (Fig. 7.18) for different values of the jerk: $\{-j_{max}, 0, +j_{max}\}$ (steps C5a, C5b and C5c in Fig. 7.15):

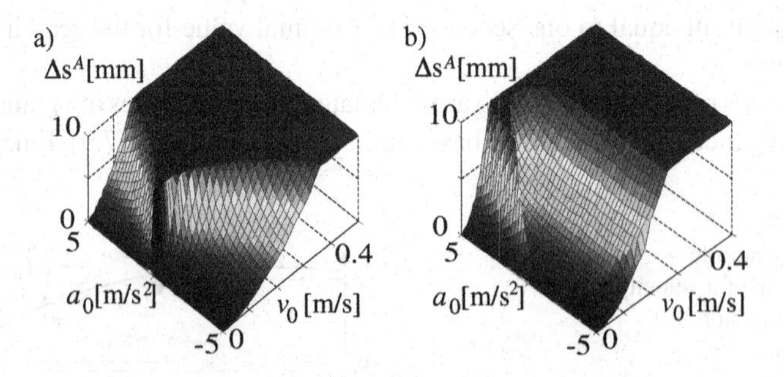

Fig. 7.18 Graphical representation of Δs^A obtained in the simulations as a function of velocity and acceleration at the given values of the jerk: **a** $j_0 = 0$, **b** $j_0 = 250$

Table 7.7 A set of the fuzzy rules of the form (4.24) of the systems of the form (4.28) used in the test trajectory validation procedure for: (a) $j_0 = -j_{max}$, (b) $j_0 = +j_{max}$

(a)	R^1: IF $(v_0$ is *high* $\vert m$ AND a_0 is *very high* $\vert m)$ THEN $(\Delta s^A$ is *very high* $\vert m)$ $\vert m$
	R^2: IF $(v_0$ is *low* $\vert m$ AND a_0 is *very low* $\vert h)$ THEN $(\Delta s^A$ is *very low* $\vert m)$ $\vert m$
	R^3: IF $(a_0$ is *high* $\vert m)$ THEN $(\Delta s^A$ is *low* $\vert m)$ $\vert l$
	R^4: IF $(v_0$ is *medium* $\vert m$ AND a_0 is *low* $\vert m)$ THEN $(\Delta s^A$ is *high* $\vert m)$ $\vert l$
(b)	R^1: IF $(a_0$ is *high* $\vert h)$ THEN $(\Delta s^A$ is *medium* $\vert m)$ $\vert l$
	R^2: IF $(v_0$ is *high* $\vert h)$ THEN $(\Delta s^A$ is *high* $\vert m)$ $\vert m$
	R^3: IF $(v_0$ is *low* $\vert l$ AND a_0 is *low* $\vert m)$ THEN $(\Delta s^A$ is *low* $\vert l)$ $\vert l$

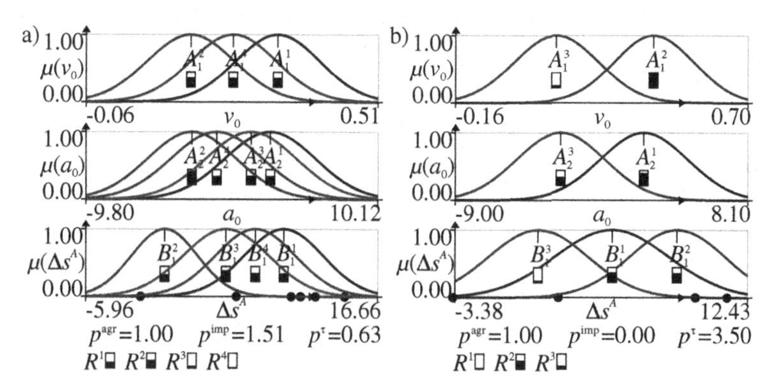

Fig. 7.19 Fuzzy sets of the fuzzy systems of the form (4.28) used in the validation procedure of the test trajectory for: **a** $j_0 = -j_{max}$, **b** $j_0 = +j_{max}$

$$\Delta s^A = \begin{cases} FS1\,(v_0, a_0) \text{ for } & r \in \{1, 7\} \\ FS2\,(v_0, a_0) \text{ for } & r \in \{2, 4, 6\} \\ FS3\,(v_0, a_0) \text{ for } & r \in \{3, 5\}, \end{cases} \tag{7.29}$$

Owing to this solution each fuzzy system used can have a simpler structure. This is an additional aspect of the transformation of the problem focused on interpretability.

- We used fuzzy systems with Gaussian membership functions of the form (4.29). Fuzzy rules of the used systems of the form (4.24) are shown in Table 7.7 and their fuzzy sets are shown in Fig. 7.19.
- The adopted approach required a declaration of the value of j_{max} at the stage of designing the control system, which is necessary before the training of the fuzzy system. This is a rather insignificant drawback because the value of j_{max} is never changed during the entire use of the machine. The selection and fixing of the value of the j_{max} as well as the initial tuning phase of the used fuzzy system must be done only once - at the stage of system designing. It is possible to prepare several fuzzy systems, each learned in advance (for different values of the j_{max} typically used in practice), and use them later without any modification. The significant

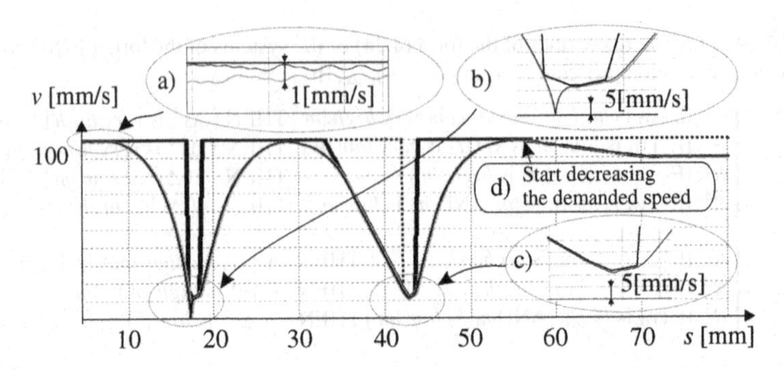

Fig. 7.20 The velocity limit (the *thick black line*), the smooth trajectory obtained with the three fuzzy systems (the *thin black line*), and the smooth trajectory obtained with the trial and error method (the *grey line*)

advantage of such approach is the simplification of the fuzzy system. As a result, the algorithm is more efficient in a real-time implementation.

- The training data, which were used in the learning process, were generated by the trial and error method. The idea of generating the training data was based on the assumption that the outputs Δs^A should have greater values, which results in decreasing the number of steps of the validation algorithm of a test trajectory. Certainly, the condition $V_E < V_{Emax}$ needs to be satisfied (Fig. 7.17).

The conclusions on the simulations can be summarized as follows:

- The final trajectory obtained with the help of the three fuzzy systems (7.29) is shown in Fig. 7.20. The comparison with the trajectory obtained by the trial and error method shows that there are some insignificant differences between them, resulting from the fuzzy systems approximation errors. Despite the slight differences between these two cases, the "fuzzy based" trajectory fully guarantees the required limits of the jerk, acceleration and velocity, and as a result it is suitable to be used in CNC systems.
- In Fig. 7.20 some areas are enlarged to better illustrate the specific features of the presented algorithm. In the first indicated area (Fig. 7.20a) the small velocity fluctuations are visible. This results from the fact that the final trajectory is formed by a merger of short fragments of the successive test trajectories. Generally, this is a drawback of the presented algorithm. However, if the amplitudes of these fluctuations are small, then it does not have a negative influence on the quality of the work. Their amplitude is proportional to the length of the connecting pieces, which actually depends on the time period T_G used to generate and validate the subsequent test trajectories. Decreasing this time period causes the reduction of the amplitude, but it requires more computational power of the computer system used. In our simulations we used T_G with an experimentally chosen value equal to 2 ms.
- The next enlarged fragment in Fig. 7.20b shows that an unfavorable slowdown occurs if the speed limit curve drops sharply. This drawback results from the lack

of the global velocity optimization techniques. However, pre-processing of the velocity limit curve, i.e. eliminating the sharp drops, could be used to prevent that adverse slowdown (Fig. 7.20c).

- Despite these minor drawbacks, a great advantage of the presented algorithm is that it is able to adjust the generated speed profile to the changing external conditions, e.g. spindle load change, in an efficient manner. In the presented approach it is possible to modify the demanded value of the feed rate of the tool during the machine operation. This is illustrated in the simulation presented in Fig. 7.20d, in which the speed limit is decreased in order to protect the spindle from overload. As we can see, the algorithm is able to modify the generated speed profile online.

7.2.4 Conclusions

In this chapter we presented the algorithm for the online speed profile generation for an industrial machine tool. A unique feature of this method is its ability to quickly adjust the generated trajectory to changing speed limits. It is possible to modify the requested value of the feed rate of the tool during the machine operation. This feature is very important for operating CNC machines because of the need to protect the cutter from the brake and spindle from overload in high-speed machining. The presented method, based on a fuzzy system, allows the algorithm to work properly and quickly, and to construct the trajectory generator operating online. It should be noted that fuzzy systems can be adopted for implementation in hardware. The algorithm discussed in this section was proposed in paper [37].

7.3 Summary

In this chapter various ways of affecting interpretability of fuzzy systems in two specific applications are discussed. We have considered an approach to weakly non-linear objects modelling and an approach to controlling of a CNC machine with a jerk limit. The basis for the developed methods included combining the possibilities of well-tested classic solutions with the possibilities offered by fuzzy systems. Although such approach requires a more complete analysis of a given problem, it seems to be promising in many different areas using to a good advantage the potential of fuzzy systems.

As already mentioned, some of the presented aspects of affecting interpretability of fuzzy systems were partially discussed in our previous papers (see e.g. [3, 31, 36, 37]).

References

1. Adjrad, M., Belouchrani, A.: Estimation of multicomponent polynomial-phase signals impinging on a multisensor array using state-space modeling. IEEE Trans. Signal Process. **55**, 32–45 (2007)
2. Banerjee, A., Arkun, Y., Ogunnaike, B., Pearson, R.: Estimation of nonlinear systems using linear multiple models. AIChE J. **43**, 1204–1226 (1997)
3. Bartczuk, Ł., Przybył, A., Cpałka, K.: A new approach to nonlinear modelling of dynamic objects based on fuzzy rules. Int. J. Appl. Math. Comput. Sci. **26**, 603–621 (2016)
4. Boukezzoula, R., Galichet, S., Foulloy, L.: Fuzzy feedback linearizing controller and its equivalence with the fuzzy nonlinear internal model control structure. Int. J. Appl. Math. Comput. Sci. **2**, 233–248 (2007)
5. Caughey, T.K.: Equivalent linearization techniques. J. Acoust. Soc. Am. **11**, 1706–1711 (1963)
6. Dieulot, J.Y., Thimoumi, I., Colas, F., Béarée, R.: Numerical aspects and performances of trajectory planning methods of flexible axes. Int. J. Comput. Commun. Control **4**, 35–44 (2006)
7. Erkorkmaz, K., Altintas, Y.: High speed CNC system design. Part I: jerk limited trajectory generation and quintic spline interpolation. Int. J. Mach. Tools Manuf. **41**, 1323–1345 (2001)
8. Grabowski, P., Callier, F.M.: Circle criterion and boundary control systems in factor form Input-output approach. Appl. Math. Comput. Sci. **11**, 1387–1403 (2001)
9. Hāber, R., Keviczky, L.: Nonlinear System Identification – Input-Output Modeling Approach. Volume 1: Nonlinear System Parameter Identification. Mathematical Modelling Theory and Applications. Kluwer, The Netherlands (1999)
10. Horzyk, A., Tadeusiewicz R.: Self-optimizing neural networks, advances in neural networks. In: Proceedings of the International Symposium on Neural Networks, pp. 150–155 (2004)
11. Huijberts, H., Nijmeijer, H., Willems, R.: System identification in communication with chaotic systems. IEEE Trans. Circ. Syst. I Fundam. Theory Appl. **47**, 800–808 (2000)
12. Ikonen, E., Najim, K.: Advanced Process Identification and Control, vol. 9. CRC Press, New York (2001)
13. Jang, I.S.R., Sun, C.T.: Neuro-fuzzy modeling and control. In: Proceedings of the IEEE, pp. 378–406 (1995)
14. Jordan, A.: Linearization of non-linear state equation. Bull. Polish Acad. Sci. Tech. Sci. **1**, 63–73 (2006)
15. Ki, N.Y.: A new velocity profile generation for high efficiency CNC machining application. City University of Hong Kong (Master Thesis), pp. 1–83 (2008)
16. Kluska, J.: Analytical Methods in Fuzzy Modeling and Control. Springer, Heidelberg (2009)
17. Kluska, J.: Selected Applications of P1-TS Fuzzy Rule-Based Systems. Lecture Notes in Computer Science, pp. 195–206. Springer, Heidelberg (2015)
18. Lei, W.T., Sung, M.P., Lin, L.Y., Huang, J.J.: Fast real-time NURBS path interpolation for CNC machine tools. Int. J. Mach. Tools Manuf. **47**, 1530–1541 (2007)
19. Lin, M.T., Tsai, M.S., Yau, H.T.: Development of an dynamics-based NURBS interpolator with real-time look-ahead algorithm. Int. J. Mach. Tools Manuf. **47**, 2246–2262 (2007)
20. Ljung, L.: Approaches to identification of nonlinear systems. In: Proceedings of the 29th Chinese Control Conference, pp. 1–5 (2010)
21. Lo Bianco, C.G., Zanasi, R.: Smooth profile generation for a tile printing machine. IEEE Trans. Ind. Electron. **50**, 471–477 (2003)
22. Macfarlane, S.: On-Line smooth trajectory planning for manipulators. The University of British Columbia (Master Thesis), pp. 1–146 (2001)
23. Mohan, S., Kweon, S.H., Lee, D.M., Yang, S.H.: Parametric NURBS curve interpolators: a review. Int. J. Precis. Eng. Manuf. **9**, 84–92 (2008)
24. Mrugalski, M.: Advanced Neural Network-Based Computational Schemes for Robust Fault Diagnosis. Studies in Computational Intelligence. Springer, Heidelberg (2014)
25. Murray-Smith, R., Johansen, T.: Multiple Model Approaches to Nonlinear Modelling and Control. CRC Press, Boca Raton (1997)

26. Nelles, O.: Nonlinear System Identification From Classical Approaches to Neural Networks and Fuzzy Models. Springer, Heidelberg (2001)
27. Ogata, K.: System Dynamics. Pearson/Prentice Hall, Upper Saddle River (2004)
28. Osario-Rios, R.A., Romero-Trosncoso, R.D.J., Herera-Ruiz, G., Casteneda-Miranda, R.: FPGA implementation of higher degree polynomial acceleration profiles for peak jerk reduction in servomotors. Robot. Comput. Integr. Manuf. **25**, 379–392 (2009)
29. Pedro, J., Dahunsi, O.: Neural network based feedback linearization control of a servo-hydraulic vehicle suspension system. Int. J. Appl. Math. Comput. Sci. **21**, 137–147 (2011)
30. Pessen, D.W.: Industrial Automation. Circuit Design and Components. Willey, New York (1989)
31. Przybył, A., Cpałka, K.: A New Method to Construct of Interpretable Models of Dynamic Systems. Lecture Notes in Computer Science, pp. 697–705. Springer, Heidelberg (2012)
32. Przybył, A., Jelonkiewicz, J.: Genetic algorithm for observer parameters tuning in sensorless induction motor drive, neural networks and soft computing. In: Proceedings of the Sixth International Conference on Neural Networks and Soft Computing. Springer, pp. 376–381 (2003)
33. Puig, V., Witczak, M., Nejjari, F., Quevedo, J., Korbicz, J.: A gmdh neural network-based approach to passive robust fault detection using a constraint satisfaction backward test. Eng. Appl. Artif. Intell. **20**, 886–897 (2007)
34. Roffel, B., Betlem, B.H.: Advanced Practical Process Control. Springer, Heidelberg (2004)
35. Rutkowski, L.: Computational Intelligence. Springer, Heidelberg (2010)
36. Rutkowski, L., Przybył, A., Cpałka, K., Er, M.J.: Online Speed Profile Generation for Industrial Machine Tool Based on Neuro Fuzzy Approach. Lecture Notes in Computer Science, vol. 6114, pp. 645–650. Springer, Heidelberg (2010)
37. Rutkowski, L., Przybył, A., Cpałka, K.: Novel on-line speed profile generation for industrial machine tool based on flexible neuro-fuzzy approximation. IEEE Trans. Ind. Electron. **59**, 1238–1247 (2012)
38. Salapa, K., Trawińska, A., Roterman, I., Tadeusiewicz, R.: Speaker identification based on artificial neural networks. Case study the polish vowel (pilot study). Bio-Algorithms Med-Syst. **10**, 91–99 (2014)
39. Schröder, D. (ed.): Intelligent Observer and Control Design for Nonlinear Systems. Springer, Heidelberg (2000)
40. Sun, Y., Wang, J., Guo, D.: Guide curve based interpolation scheme of parametric curves for precision CNC machining, parametric NURBS curve interpolators: a review. Int. J. Mach. Tools Manuf. **46**, 235–242 (2006)
41. Sun, F., Li, L., Li, H.X., Liu, H.: Neuro-fuzzy dynamic-inversion-based adaptive control for robotic manipulators-discrete time case. IEEE Trans. Ind. Electron. **54**, 342–1351 (2007)
42. Tadeusiewicz, R.: Using neural networks for simplified discovery of some psychological phenomena. Artif. Intell. Soft Comput. **6114**, 104–123 (2010)
43. Tadeusiewicz, R., Figura, I.: Phenomenon of tolerance to damage in artificial neural networks. Comput. Methods Mater. Sci. **11**, 501–513 (2011)
44. Tadeusiewicz, R., Chaki, R., Chaki, N.: Exploring Neural Networks with C#. CRC Press, Boca Raton (2014)
45. Takagi, T., Sugeno, M.: Fuzzy identification of systems and its application to modeling and control. IEEE Trans. Syst. Man Cybern. **15**, 116–132 (1985)
46. Tan, Z.: Time-varying time-delay estimation for nonlinear systems using neural networks. Int. J. Appl. Math. Comput. Sci. **1**, 63–68 (2004)
47. Tsai, M.S., Nien, H.W., Yau, H.T.: Development of an integrated look-ahead dynamics-based NURBS interpolator for high precision machinery. Comput. Aided Des. **40**, 554–566 (2008)
48. Tseng, S.J., Lin, K.Y., Lai, J.Y., Ueng, W.D.: A NURBS curve interpolator with jerk-limited trajectory planning. J. Chin. Inst. Eng. **32**, 215–228 (2009)
49. Witkowska, A., Śmierzchalski, R.: Designing a ship course controller by applying the adaptive backstepping method. Int. J. Appl. Math. Comput. Sci. **4**, 985–997 (2012)
50. Xie, Y., Guo, B., Xu, L., Li, J., Stoica, P.: Multistatic adaptive microwave imaging for early breast cancer detection. IEEE Trans. Biomed. Eng. BME **53**, 1647–1657 (2006)

51. Yau, H.T., Wang, J.B.: Fast Bezier interpolator with real-time lookahead function for high-accuracy machining. Int. J. Mach. Tools Manuf. **47**, 1518–1529 (2007)
52. Yau, H.T., Wang, J.B., Hsu, C.Y., Yeh, C.H.: PC-based controller with real-time look-ahead NURBS interpolator. Comput. Aided Des. Appl. **4**, 331–340 (2007)
53. Zheng, C., Su, Y., Müller, P.C.: Simple on-line smooth trajectory generation for industrial systems. Mechatronics **19**, 571–576 (2009)

Chapter 8
Case Study: Interpretability of Fuzzy Systems Applied to Identity Verification

In Sect. 2.2 and Chaps. 5–7 we present typical schemes of fuzzy system learning, which can be used, for example, in the case when we do not have expert knowledge but we have learning data. In practice, solutions are also used in which an expert initially determines the number of rules and approximate distribution of fuzzy sets with the aim of a learning process being the fine tuning of a fuzzy system. The solutions proposed in this chapter are used to improve interpretability of fuzzy systems by eliminating the supervised learning process. Their idea is an appropriate initialization of fuzzy system rules which relies on various factors including analytical determination of values of parameters determining fuzzy sets localization. It is usually possible when physical meaning of these parameters can be interpreted.

In this chapter we describe methods of improving fuzzy systems interpretability by elimination of the supervised learning in biometric issues with a special emphasis on the issue of identity verification on the basis of dynamic features of a handwritten electronic signature. We describe two different algorithms (Sects. 8.1 and 8.2). We present the simulation results (Sects. 8.1.2 and 8.2.2) and conclusions (Sects. 8.1.3 and 8.2.3) for both algorithms. Moreover, the final part of the chapter contains its summary (Sect. 8.3) and bibliography.

8.1 Identity Verification on the Basis of Characteristic Partitions of the Dynamic Signature

Security of IT systems is related to a number of factors including effective identity verification of system users. This verification can be performed using various methods based on: (a) something you have (e.g. chip card), (b) something you know (e.g. password), or (c) something you are (e.g. biometric features). The third approach is the most convenient for people whose identity is being verified and the most

K. Cpałka, *Design of Interpretable Fuzzy Systems*, Studies in Computational Intelligence 684, DOI 10.1007/978-3-319-52881-6_8

difficult to forge for potential forgers. Therefore, it is rather compelling and it creates possibilities for the development of new solutions. The biometric features used in this approach are divided into two categories: (a) physiological ones - related to the construction of the human body (e.g. fingerprint, iris, hand geometry, face) and (b) behavioral ones - related to the human behavior (e.g. signature, gait, keystrokes). A handwritten signature occupies a special place among behavioral characteristics, its acquisition is not controversial and it is commonly socially acceptable.

In the literature there are two main approaches to signature analysis. The first one uses so-called static (off-line) signature and it is based on an analysis of geometric features of the signature, such as shape and size ratios, etc. [2–4, 17, 26, 41]. The other approach is based on an analysis of the dynamics of a signing process and it uses so-called dynamic (online) signature. Some authors have also presented methods based on these both approaches [15, 40]. The most commonly used signals, which are the basis of the dynamic signature analysis, are pen pressure on the tablet surface and pen velocity. The velocity signal is determined indirectly on the basis of pen position signals on the tablet surface. The dynamic signature verification is much more effective than the static one because: (a) dynamics of signing is a very individual characteristic feature of the signer, (b) it is difficult to forge, (c) waveforms describing the dynamics of the signature (even if you have them) are difficult to translate into the process of signing, but they are relatively easy to analyze. It should also be emphasized that the algorithms for the analysis of the dynamic signature can be relatively easily used in other areas of application in the field of biometrics, which are based on the analysis of dynamic behavior [8, 10].

In the literature there are four main approaches to the dynamic signature analysis: (a) global feature-based approach [14, 29, 32–34], (b) function-based approach [11, 21, 23, 25, 30, 35], (c) regional approach [5, 6, 12, 13, 19, 24, 39, 44] and (d) hybrid approach [31, 36, 38]. Among these approaches to the dynamic signature analysis, the methods based on signature regions are very interesting and effective. In the literature in this field one can find, for example, new methods of selection of signature regions characteristic of the signer, new ways of interpretation of these signature regions and new ways of signature classification based on selected regions. Many authors use Hidden Markov Models [13]. Other authors propose ways of classification adapted to their methods. In [19] the signatures are segmented into strokes and for each of them the reliability measure is computed on the basis of the feature values which belong to the current stroke. In [24] a stroke-based algorithm that splits the velocity signal into three bands has been proposed. This approach assumes that low and high-velocity bands of the signal are unstable, whereas the medium-velocity band can be used for discrimination purposes. A more detailed review of the literature on the dynamic signature verification has been presented in e.g. [5, 6].

The method discussed in this section uses the fuzzy system of the form (4.28) in the verification process. Its structure and operation are consistent for all users. However, parameters of the system are selected individually for each user. The interpretability criteria described in Sects. 3.4, 4.4 and 6.3 are taken into account during the construction of the system. Other features of this method can be summarized as follows:

- It selects the partitions of the signature which have the following interpretation: high and low velocity in the initial, middle and final time moments of signing, high and low pressure in the initial, middle and final time moments of signing.
- It determines values of weights of importance for each partition. Weights values are proportional to the stability of reference signatures in the partitions. Thanks to this, the method uses all the partitions in the evaluation of signature similarity (with varying intensity).
- It bases on four types of signals: a shape signal of the trajectory x, a shape signal of the trajectory y, a pressure signal of the pen z and a velocity signal of the pen v. They are available as a standard for graphics tablets: the first three of them are acquired directly from the graphics tablet and the velocity is the first derivative of a signature trajectory. Various types of tablets may have different sampling frequency, so, in this case, acquired signals are subject to the standard normalization procedure. In addition, signatures should be pre-processed using other standard methods to match their length, rotation, scale and offset.
- It does not require so-called skilled forgeries and reference signatures of other signers in the training phase (which is a big advantage in this group of methods). It is possible because learning of a fuzzy system used for signature verification is not necessary (as it is a one-class classifier).
- It allows to flexibly adjust a set of signals describing the dynamics of the signature to specific areas of application and hardware capabilities. There are two most common variants of the method. The first one assumes that a graphics tablet is used in the training phase and in the verification phase. In this case, the precision of the presented method is the highest, because the signals describing not only the shape of the signature, but also its dynamics, are used in both phases. The second variant assumes that in the training phase the graphics tablet is available and in the verification phase we have a stand-alone (not connected to the computer) device with a touch screen (e.g. a smartphone, a tablet, etc.), from which it is impossible to obtain information about the pen pressure. In this case, the partitions are determined in the training phase on the basis of velocity and pressure. The verification phase takes into account only the shape of the signature. In the description of the method we took into account the second variant, assuming that in the signature verification phase the signals describing the dynamics of the signature may not be available. Obviously, in practice, there may be also indirect modes of action (e.g. based on the generation of velocity trajectories in the training phase without knowing the pressure trajectory or using the angle of the pen to the tablet surface during the signing process), but this method can be adapted to each of them.

8.1.1 Description of the Method

The algorithm for the dynamic signature verification based on hybrid partitioning works in two phases (see Fig. 8.1): the training phase (Sect. 8.1.1.1) and the test phase

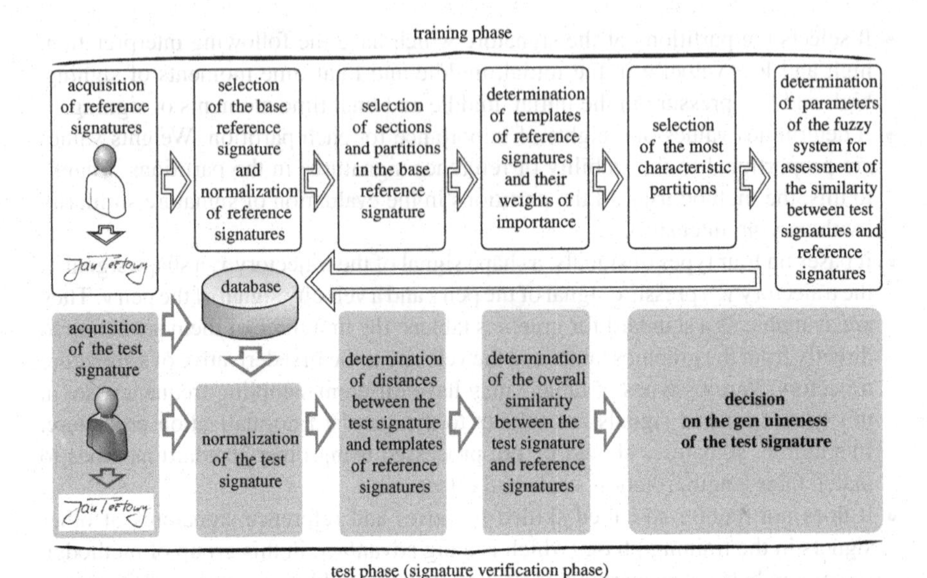

Fig. 8.1 A block diagram of the algorithm (discussed in this section) for the dynamic signature verification based on hybrid partitioning

(Sect. 8.1.1.2). In both of them a procedure of signature normalization is performed (see Fig. 8.2). In this procedure the most typical reference signature, called base signature, is selected for each user. It is one of the reference signatures collected in the acquisition phase, for which the distance to the other reference signatures is the smallest. The distance is calculated according to the adopted (e.g. Euclidean) distance measure. Training or test signatures are matched to the base signature using the Dynamic Time Warping (DTW) algorithm [1, 9, 27], which operates on the basis of matching velocity and pressure signals. The result of matching the two signatures is a map of their corresponding points. On the basis of the map, trajectories of the signatures are matched. Matching by way of using the DTW algorithm could not be done directly with the use of trajectories, because this would remove the differences between the shapes of the signatures. It would have a very negative impact on training. Elimination of differences in rotation of the signatures is performed by the Principal Component Analysis (PCA) algorithm which, in the literature, is commonly used to make image rotation invariant [16]. The scale and offset are compensated by standard geometric transformations. Various normalization techniques are described in detail in the literature [20, 21, 28, 37].

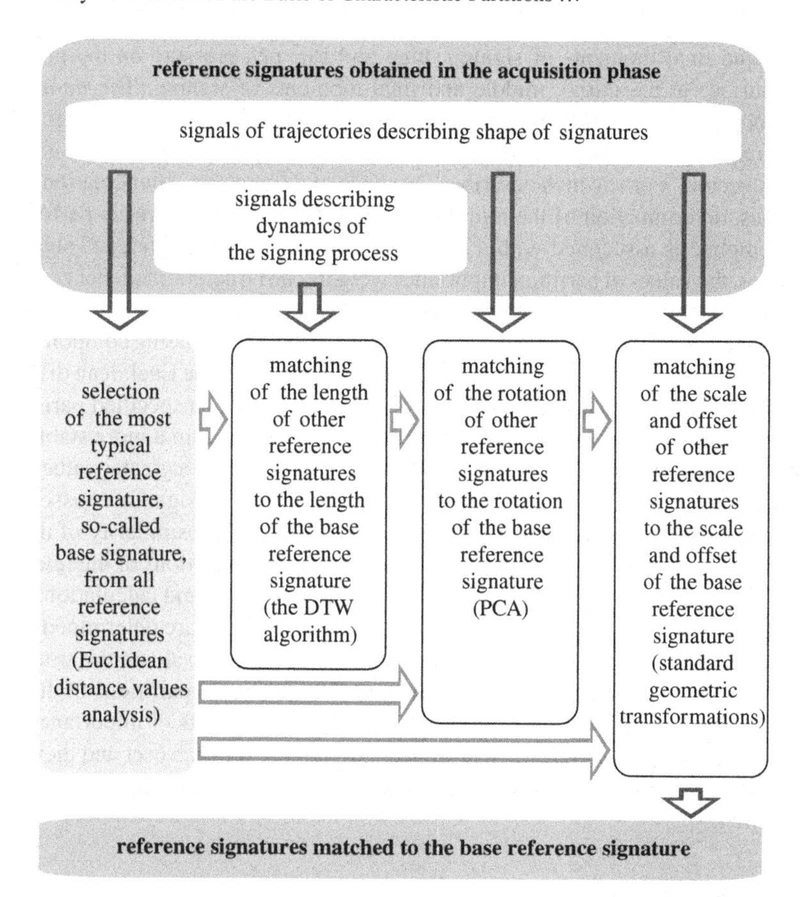

Fig. 8.2 Signature normalization procedure

8.1.1.1 Training Phase

At the beginning of the training phase partitioning of the reference signatures of each user is carried out. They are hybrid partitions because they are created from a combination of vertical and horizontal sections. Vertical sections are time intervals indicating the initial, middle and final phases of the signing process. Horizontal sections are created in each vertical section. In this process signals describing the dynamics of a signature are taken into account. If the velocity signal is partitioned first (order of signal processing is arbitrary), then in each time interval the average value of the velocity is determined (averaging discrete values of the velocity). Next, in each time interval two partitions are created: (a) the one associated with the velocity lower than the average and (b) the other one associated with the velocity higher than the average. The procedure is analogous for the second available signal – the pen pressure. As a result, the following partitions are created: high and low velocity at the initial, middle and final moments of signing, high and low pen pressure on the graphics tablet surface at the initial, middle and final moments of signing. The number of time moments which affects the number of partitions is a parameter of the algorithm, which may be greater than or equal to 1. Each partition is physically a subset of points of the trajectory x or y, which describes the shape of a signature. After creation of the partitions, determination of the templates of the reference signatures is performed. Each template is associated with a separate partition. With the reference signature templates, the values of partition importance weights can be determined. For example, the weight of importance of the first partition depends on the similarity of the points of the reference signatures from the first partition to the corresponding components of the first template. This similarity is dependent on the value of the Euclidean distance. A higher value of the partition importance weight means that a specified part of the reference signatures associated with this partition was created in a more stable way (with a similar value of the velocity or the pressure). Moreover, a greater value of the weight means that in the test phase the fragments of the test signatures associated with this partition will be more important in evaluation of the similarity of the test signature to the reference signatures. After creation of the partitions of the reference base signature, creation of the reference signatures templates and calculation of the weights of importance, parameters of the fuzzy system (4.28) are determined in the learning phase. The fuzzy system is used in the test phase to evaluate the similarity of the test signature to the reference signatures. Obviously, the partitions of the base signature, the templates of the reference signatures, the weights of importance, and the parameters of the system are determined individually for each user and they must be stored in a database. It is the most important feature of the method presented in this section.

Creating Partitions

Each reference signature j ($j = 1, 2, \ldots, J$, where J is the number of reference signatures) of the user i ($i = 1, 2, \ldots, I$, where I is the number of users) is represented by the following signals:

- **Signals describing the shape of a signature**. Signal $\mathbf{x}_{i,j} = \left[x_{i,j,k=1}, x_{i,j,k=2}, \ldots, x_{i,j,k=K_i}\right]$ describes the movement of the pen in the two-dimensional space along the x axis, where K_i is the number of signal samples. The number of samples K_i results from the sampling frequency of the graphics tablet and performance of the DTW algorithm in the normalization phase. Thanks to the signature normalization, all trajectories describing the signatures of the user i have the same number of samples K_i. Movement of the pen along the y axis can be described in a similar way: $\mathbf{y}_{i,j} = \left[y_{i,j,k=1}, y_{i,j,k=2}, \ldots, y_{i,j,k=K_i}\right]$. In order to simplify the description of the algorithm we used the same symbol $\mathbf{a}_{i,j} = \left[a_{i,j,k=1}, a_{i,j,k=2}, \ldots, a_{i,j,k=K_i}\right]$ to describe both shape trajectories, where $a \in \{x, y\}$.
- **Signals describing the dynamics of a signature**. Signal $\mathbf{v}_{i,j} = \left[v_{i,j,k=1}, v_{i,j,k=2}, \ldots, v_{i,j,k=K_i}\right]$ describes the velocity of the pen and trajectory $\mathbf{z}_{i,j} = \left[z_{i,j,k=1}, z_{i,j,k=2}, \ldots, z_{i,j,k=K_i}\right]$ describes the pen pressure on the surface of a graphics tablet.

In order to simplify the description of the algorithm we used the same symbol $\mathbf{s}_{i,j} = \left[s_{i,j,k=1}, s_{i,j,k=2}, \ldots, s_{i,j,k=K_i}\right]$ to describe both dynamics signals, where $s \in \{v, z\}$.

The purpose of the partitioning is to assign each point of the signal $\mathbf{v}_{i,jBase}$ and the signal $\mathbf{z}_{i,jBase}$ of the reference base signature to a single hybrid partition resulting from a combination of the vertical and the horizontal sections, where $jBase \in \{1, \ldots, J\}$ is an index of the base signature. As already mentioned, the base signature is taken into account during partitioning as the most typical reference signature of the user i. Therefore, it is selected from a set of reference signatures and it is not generated by averaging or grouping signals describing reference signatures.

The idea of partitioning is shown in Fig. 8.3. At the beginning of the partitioning, the vertical sections of the signals $\mathbf{v}_{i,jBase}$ and $\mathbf{z}_{i,jBase}$ are created. Each of them

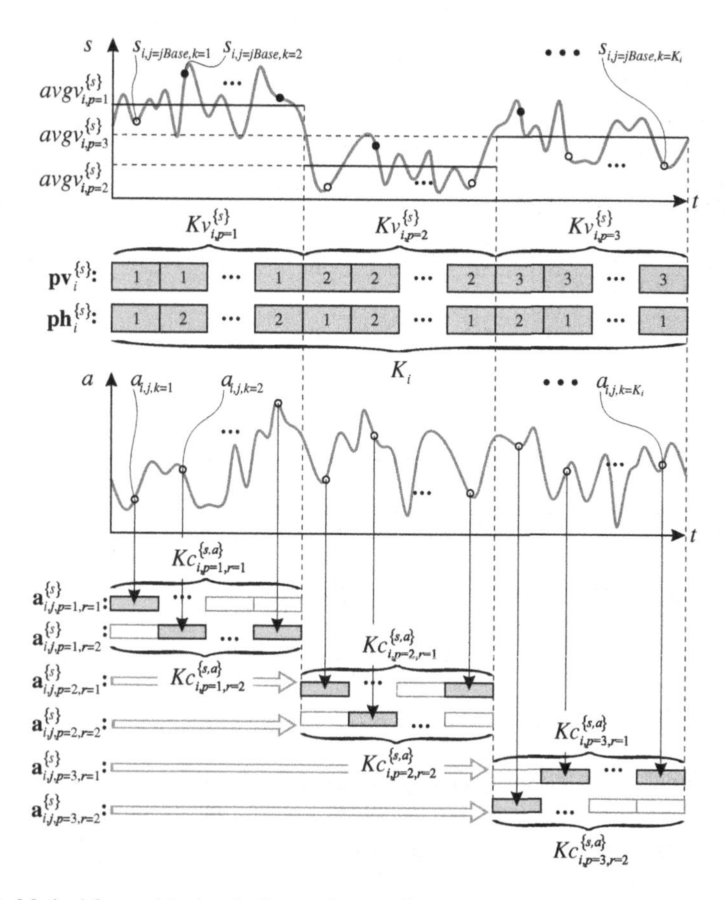

Fig. 8.3 Method for partitioning the base reference signature $j = jBase$ on the basis of the signal $s \in \{v, z\}$ and an example of partitioning the trajectory $a \in \{x, y\}$ of the reference signature j of the user i for the case $P^{\{s\}} = 3$ and $R^{\{s\}} = 2$. The method determining sections for different combination of the values $P^{\{s\}}$ and $R^{\{s\}}$ is conducted in a similar way

represents a different time moment of signing: (a) initial or final for the case $P^{\{s\}} = 2$, (b) initial, middle or final for the case $P^{\{s\}} = 3$, (c) initial, first middle, second middle or final for the case $P^{\{s\}} = 4$. The vertical sections are indicated by the elements of the vector $\mathbf{pv}_i^{\{s\}} = \left[pv_{i,k=1}^{\{s\}}, pv_{i,k=2}^{\{s\}}, \ldots, pv_{i,k=K_i}^{\{s\}}\right]$ determined as follows:

$$pv_{i,k}^{\{s\}} = \begin{cases} 1 \text{ for } & 0 < k \leq \frac{K_i}{P^{\{s\}}} \\ 2 \text{ for } & \frac{K_i}{P^{\{s\}}} < k \leq \frac{2K_i}{P^{\{s\}}} \\ \vdots & \\ P^{\{s\}} \text{ for } & \frac{(P^{\{s\}}-1)K_i}{P^{\{s\}}} < k \leq K_i, \end{cases} \tag{8.1}$$

where $s \in \{v, z\}$ is the signal type used for determination of the partition (velocity v or pressure z), i is the user index ($i = 1, 2, \ldots, I$), j is the reference signature index ($j = 1, 2, \ldots, J$), K_i is the number of samples of normalized signals of the user i (divisible by $P^{\{s\}}$), k is an index of the signal sample ($k = 1, 2, \ldots, K_i$) and $P^{\{s\}}$ is the number of the vertical sections ($P^{\{s\}} \ll K_i$ and $P^{\{s\}} = P^{\{v\}} = P^{\{z\}}$). The number of vertical sections can be arbitrary, but its increase does not increase interpretability and accuracy of the method.

After the creation of the vertical sections of the signals $\mathbf{v}_{i,jBase}$ and $\mathbf{z}_{i,jBase}$, their horizontal sections are created. Each of them represents high and low velocity and high and low pressure at individual moments of signing. The horizontal sections indicated by the elements of the vector $\mathbf{ph}_i^{\{s\}} = \left[ph_{i,k=1}^{\{s\}}, ph_{i,k=2}^{\{s\}}, \ldots, ph_{i,k=K_i}^{\{s\}}\right]$ are determined as follows:

$$ph_{i,k}^{\{s\}} = \begin{cases} 1 \text{ for } s_{i,j=jBase,k} < avgv_{i,p=pv_{i,k}^{\{s\}}}^{\{s\}} \\ 2 \text{ for } s_{i,j=jBase,k} \geq avgv_{i,p=pv_{i,k}^{\{s\}}}^{\{s\}}, \end{cases} \tag{8.2}$$

where $jBase$ is the base signature index, $avgv_{i,p}^{\{s\}}$ is an average velocity (when $s = v$) or an average pressure (when $s = z$) in the section indicated by the index p of the base signature $jBase$:

$$avgv_{i,p}^{\{s\}} = \frac{1}{Kv_{i,p}} \sum_{k=\left(\frac{(p-1)\cdot K_i}{P^{\{s\}}}+1\right)}^{k=\left(\frac{p\cdot K_i}{P^{\{s\}}}\right)} s_{i,j=jBase,k}, \tag{8.3}$$

where $Kv_{i,p}$ is the number of samples in the vertical section p, $s_{i,j=jBase,k}$ is the sample k of the signal $s \in \{v, z\}$ describing dynamics of the signature.

As a result of partitioning, each sample $v_{i,jBase,k}$ of the signal $\mathbf{v}_{i,jBase}$ of the base signature $jBase$ and each sample $z_{i,jBase,k}$ of the signal $\mathbf{z}_{i,jBase}$ of the base signature $jBase$ is assigned to the vertical section (assignment information is stored in the vector $\mathbf{pv}_i^{\{s\}}$) and the horizontal section (assignment information is stored in the vector $\mathbf{ph}_i^{\{s\}}$). The intersection of the sections creates the partition. The fragments of the shape trajectories $\mathbf{x}_{i,j}$ and $\mathbf{y}_{i,j}$, created by taking into account $\mathbf{pv}_i^{\{s\}}$ and $\mathbf{ph}_i^{\{s\}}$, will be denoted as $\mathbf{a}^{\{s\}}_{i,j,p,r} = \left[a^{\{s\}}_{i,j,p,r,k=1}, a^{\{s\}}_{i,j,p,r,k=2}, \ldots, a^{\{s\}}_{i,j,p,r,k=Kc^{\{s,a\}}_{i,p,r}} \right]$. The number of samples belonging to the partition (p, r) (created as an intersection of the vertical section p and the horizontal section r, included in the trajectory $\mathbf{a}^{\{s\}}_{i,j,p,r}$) of the user i associated with the signal a (x or y) and created on the basis of the signal s (velocity or pressure) will be denoted as $Kc^{\{s,a\}}_{i,p,r}$. It should be noticed that $\sum_{p=1}^{P^{\{s\}}} \sum_{r=1}^{R^{\{s\}}} Kc^{\{s,a\}}_{i,p,r} = K_i$ for $a \in \{x, y\}$ and $s \in \{v, z\}$. The number of partitions of the base signature of the user i is equal to $P^{\{v\}} \cdot P^{\{z\}} \cdot 4$. The partitions are used to determine the reference signature templates.

Generating Templates

The following parameters are considered when determining reference signature templates: (a) all J reference signatures of the user i, (b) two shape trajectories of the reference signatures, i.e. $\mathbf{x}_{i,j}$ and $\mathbf{y}_{i,j}$ and (c) partitions created for the reference base signature, resulting from the intersection of the vertical sections (indicated by $\mathbf{pv}^{\{s\}}_i$) and horizontal sections (indicated by $\mathbf{ph}^{\{s\}}_i$) (Fig. 8.4). The templates of the signatures are averaged fragments of the reference signatures represented by the shape trajectories $\mathbf{x}_{i,j}$ or $\mathbf{y}_{i,j}$. The number of the templates created for the user i is equal to the number of the partitions. Each template $\mathbf{tc}^{\{s,a\}}_{i,p,r} = \left[tc^{\{s,a\}}_{i,p,r,k=1}, tc^{\{s,a\}}_{i,p,r,k=2}, \ldots, tc^{\{s,a\}}_{i,p,r,k=Kc^{\{s,a\}}_{i,p,r}} \right]$ describes fragments of the reference signatures in the partition (p, r) of the user i, associated with the signal a (x or y), created on the basis of the signal s (velocity or pressure), where:

$$tc^{\{s,a\}}_{i,p,r,k} = \frac{1}{J} \sum_{j=1}^{J} a^{\{s\}}_{i,j,p,r,k}. \tag{8.4}$$

After the templates $\mathbf{tc}^{\{s,a\}}_{i,p,r}$ are generated, parameters of the fuzzy system (4.28) for evaluating the similarity of the test signatures to the reference signatures are determined.

Determining the Parameters of the Fuzzy System to Evaluate the Similarity of the Test Signatures to the Reference Signatures

The test signature verification is based on the answers of the fuzzy system (4.28) for evaluating the similarity of the test signatures to the reference signatures. Parameters of the system must be selected individually for each user from the database. Moreover, the algorithm for signature verification should: (a) work independently of the number of users (its accuracy should not depend on the number of users in the database), (b) have the ability to easily add signatures of new users, (c) not take into account signatures of other users in the training and verification phases of the signature. This

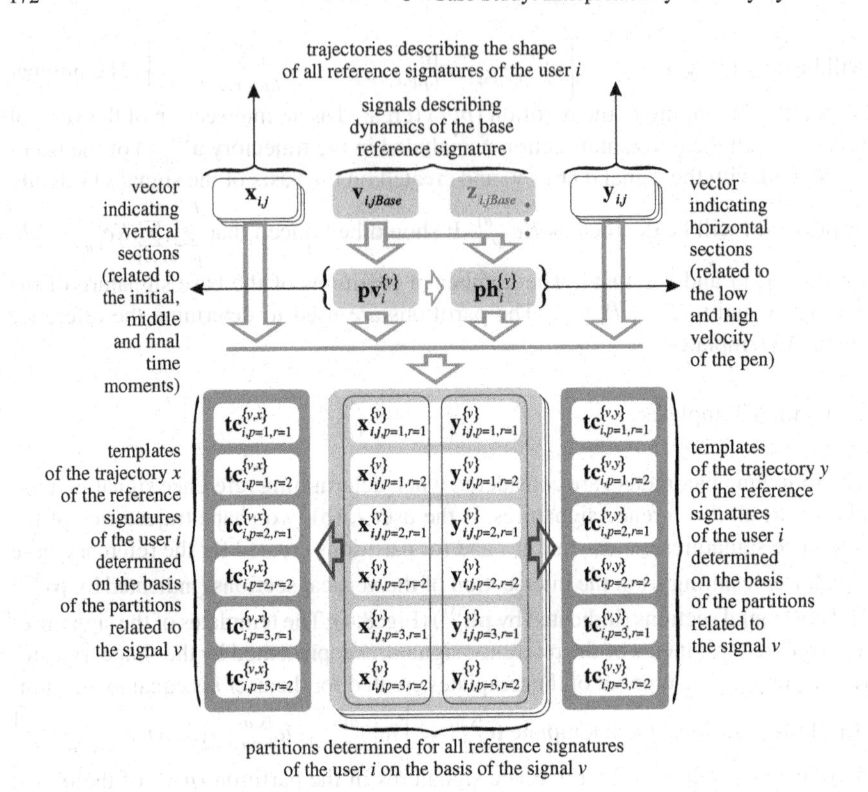

Fig. 8.4 Determining trajectories x and y of the templates of the user i on the basis of the partitions associated with the signal v for $P^{\{s\}} = 3$ and $R^{\{s\}} = 2$. Determining the templates on the basis of the partitions associated with the signal z and different values of parameters $P^{\{s\}}$ and $R^{\{s\}}$ is done in a similar way

limits the use of the known methods e.g. from the field of nonlinear classification and machine learning (both evolutionary or gradient). The algorithm considered in this section has these features, which results from the appropriate use of the fuzzy system of the form (4.28).

The first group of parameters of the considered system are the parameters describing differences between the reference signatures and the templates in the partitions. They are used in the construction of fuzzy rules in the form (4.24) and are determined as follows:

$$dmax_{i,p,r}^{\{s,a\}} = \frac{\delta_i}{J \cdot Kc_{i,p,r}^{\{s,a\}}} \cdot \sum_{k=1}^{Kc_{i,p,r}^{\{s,a\}}} \sum_{j=1}^{J} \left| a_{i,j,p,r,k}^{\{s\}} - tc_{i,p,r,k}^{\{s,a\}} \right|, \tag{8.5}$$

where δ_i is a parameter which ensures matching of tolerance of the system for evaluating similarity in the test phase (it is assumed that the test signatures can be created in less comfortable conditions than the reference signatures, thus $\delta_i \geq 1$). Thus, values of the parameters describing differences between the reference signatures and the templates in the partitions of the user i take into account the average similarity of the reference signature shape to the templates in the partitions.

The second group of parameters of the considered system are weights of the partitions. They are used for evaluation of the similarity of the test signatures to the templates of the reference signatures of the user i. A high value of the weight means a small dispersion of the shape signal values from the template $\mathbf{tc}_{i,p,r}^{\{s,a\}}$ for the reference signatures of the user i. A consequence of the high value of the partition weight is lower tolerance of the system for similarity evaluation in the test phase.

Determination of the weights $w_{i,p,r}^{\{s,a\}}$ starts with determination of a dispersion of the reference signatures signals. The dispersion is represented by a standard deviation. The average standard deviation for all samples in the partition is determined as follows:

$$\bar{\sigma}_{i,p,r}^{\{s,a\}} = \frac{1}{Kc_{i,p,r}^{\{s,a\}}} \cdot \sum_{k=1}^{Kc_{i,p,r}^{\{s,a\}}} \sqrt{\frac{1}{J} \sum_{j=1}^{J} \left(a_{i,j,p,r,k}^{\{s\}} - tc_{i,p,r,k}^{\{s,a\}} \right)^2}. \tag{8.6}$$

With the average standard deviation $\bar{\sigma}_{i,p,r}^{\{s,a\}}$, normalized values of the partition weights are determined:

$$w_{i,p,r}^{\{s,a\}} = 1 - \frac{\bar{\sigma}_{i,p,r}^{\{s,a\}}}{\max\limits_{\substack{p=1,2,\ldots,P^{\{s\}} \\ r=1,2}} \left\{ \bar{\sigma}_{i,p,r}^{\{s,a\}} \right\}}. \tag{8.7}$$

Normalization of the weights adapts them for use in the fuzzy system (4.28) used for evaluation of the similarity of the test signatures to the reference signatures. This evaluation is the basis for recognition of signature authenticity.

8.1.1.2 Test Phase (Verification of Signatures)

At the beginning of the test phase (Fig. 8.1) the user: (a) creates one signature, which will be verified and (b) claims to be a specific user from the database. Then, the parameters of this particular user, stored earlier in the database, are downloaded and the signature verification is performed. The list of the parameters is as follows: (a) trajectories of the base signature $\mathbf{x}_{i,jBase}$, $\mathbf{y}_{i,jBase}$, $\mathbf{v}_{i,jBase}$ and $\mathbf{z}_{i,jBase}$, (b) vectors of allocation to the sections $\mathbf{pv}_i^{\{s\}}$ and $\mathbf{ph}_i^{\{s\}}$, (c) templates of the reference signatures $\mathbf{tc}_{i,p,r}^{\{s,a\}}$, (d) weights of the partitions $w_{i,p,r}^{\{s,a\}}$ ($p = 1, 2, \ldots, P^{\{s\}}$, $r = 1, 2$), and (e) the parameters describing differences between the reference signatures and the templates in the partitions $dmax_{i,p,r}^{\{s,a\}}$.

The first step of the verification phase is acquisition of the test signature. This signature is pre-matched to the reference base signature, represented by the trajectories $\mathbf{x}_{i,jBase}$, $\mathbf{y}_{i,jBase}$ and the signals $\mathbf{v}_{i,jBase}$, $\mathbf{z}_{i,jBase}$. This is done in a similar way as in the case of the reference signatures in the training phase (Fig. 8.2). The normalized test signature is represented by two shape trajectories: $\mathbf{xtst}_i = \left[xtst_{i,k=1}, xtst_{i,k=2}, \ldots, xtst_{i,k=K_i}\right]$ and $\mathbf{ytst}_i = \left[ytst_{i,k=1}, ytst_{i,k=2}, \ldots, ytst_{i,k=K_i}\right]$. Their structure is analogous to the shape trajectories $\mathbf{x}_{i,j}$ and $\mathbf{y}_{i,j}$ of the reference signatures used in the training phase, but they do not have the index j pointing to the signature.

The second step of the verification phase is partitioning of the test signature. As a result of partitioning of the shape trajectories $\mathbf{xtst}_i$ and $\mathbf{ytst}_i$ their fragments denoted as $\mathbf{atst}_{i,p,r}^{\{s\}} = \left[a_{i,p,r,k=1}^{\{s\}}, a_{i,p,r,k=2}^{\{s\}}, \ldots, a_{i,p,r,k=Kc_{i,p,r}^{\{s,a\}}}^{\{s\}}\right]$ are obtained. During the partitioning the vectors $\mathbf{pv}_i^{\{s\}}$ and $\mathbf{ph}_i^{\{s\}}$ are used. They are determined in the training phase and their signals indicate sections, in which signals of the vectors $\mathbf{xtst}_i$ and $\mathbf{ytst}_i$ should be placed. It was carried out in a similar way in the training phase.

The third step of the verification phase (realized after partitioning) is determination of the similarity of fragments of the test signature shape trajectories $\mathbf{dtst}_{i,p,r}^{\{s\}}$ to the templates of the reference signatures $\mathbf{tc}_{i,p,r}^{\{s,a\}}$ in the partition (p, r) of the user i associated with the signal a (x or y) created on the basis of the signal s (velocity or pressure). This is determined as follows:

$$dtst_{i,p,r}^{\{s,a\}} = \frac{1}{Kc_{i,p,r}^{\{s,a\}}} \cdot \sum_{k=1}^{Kc_{i,p,r}^{\{s,a\}}} \left| atst_{i,p,r,k}^{\{s\}} - tc_{i,p,r,k}^{\{s,a\}} \right|. \tag{8.8}$$

After determination of the similarities $dtst_{i,p,r}^{\{s,a\}}$, the total similarity of the test signature to the reference signatures of the user i is determined. A decision on the authenticity of the test signature is taken on the basis of this similarity.

Evaluation of the Overall Similarity of the Test Signature to the Reference Signatures

The fuzzy system (4.28) evaluating similarity of the test signature to the reference signatures works on the basis of the signals $dtst_{i,p,r}^{\{s,a\}}$ and takes into account the weights $w_{i,p,r}^{\{s,a\}}$. Its response is the basis for the evaluation of the signature reliability. The considered system works on the basis of two fuzzy rules (4.24) presented as follows:

$$\left\{\begin{array}{l} R^1: \left[\left(\begin{array}{c} \text{IF}\left(dtst_{i,1,1}^{\{v,x\}} \text{ is } A_{i,1,1}^{1\{v,x\}}\right)\Big|w_{i,1,1}^{\{v,x\}} \text{ AND } \ldots \\ \ldots \text{ AND } \left(dtst_{i,P^{\{z\}},2}^{\{z,y\}} \text{ is } A_{i,P^{\{z\}},2}^{1\{z,y\}}\right)\Big|w_{i,P^{\{z\}},2}^{\{z,y\}} \\ \text{THEN } \left(y_i \text{ is } B^1\right)|1 \end{array}\right)\Big|1\right] \\ R^2: \left[\left(\begin{array}{c} \text{IF}\left(dtst_{i,1,1}^{\{v,x\}} \text{ is } A_{i,1,1}^{2\{v,x\}}\right)\Big|w_{i,1,1}^{\{v,x\}} \text{ AND } \ldots \\ \ldots \text{ AND } \left(dtst_{i,P^{\{z\}},2}^{\{z,y\}} \text{ is } A_{i,P^{\{z\}},2}^{2\{z,y\}}\right)\Big|w_{i,P^{\{z\}},2}^{\{z,y\}} \\ \text{THEN } \left(y_i \text{ is } B^2\right)|1 \end{array}\right)\Big|1\right], \end{array}\right. \tag{8.9}$$

where

- $dtst_{i,p,r}^{\{s,a\}}$ ($i = 1, 2, \ldots, I$, $p = 1, 2, \ldots, P^{\{s\}}$, $r = 1, 2$, $s \in \{v, z\}$, $a \in \{x, y\}$) are input linguistic variables describing similarity of the shape trajectories' fragments $\mathbf{atst}_{i,p,r}^{\{s\}}$ of the test signature to the templates of the reference signatures $\mathbf{tc}_{i,p,r}^{\{s,a\}}$. "High" and "low" values taken by these variables are Gaussian fuzzy sets $A_{i,p,r}^{1\{v,x\}}$, $A_{i,p,r}^{2\{v,x\}}$ (see Fig. 8.5) described by the formula (4.29). In the used system the value of the center of the Gaussian function for the rule R^1 from the rule base (8.9) is equal to 0 and for the rule R^2 is equal to the value of the border of inclusion of the reference signatures $dmax_{i,p,r}^{\{s,a\}}$ calculated by the formula (8.5). The value of the width of the Gaussian function (σ) for both rules from the rule base (8.9) is determined as follows:

$$\sigma = \frac{dmax_{i,p,r}^{\{s,a\}}}{\sqrt{|\log\left(\mu_{\min}\right)|}}, \tag{8.10}$$

  where $\mu_{\min} > 0$ is a small positive number resulting from the intersection of the Gaussian function (4.29) with a straight line, described by the equation $\mu(x) = \mu_{\min}$, at the point $\left(dmax_{i,p,r}^{\{s,a\}}, \mu_{\min}\right)$. This approach results from the specificity of the Gaussian function, which tends asymptotically to the value 0 (this is the case in which $\mu_{\min} = 0$), but it never reaches it.
- y_i ($i = 1, \ldots, I$) is the output linguistic variable meaning "similarity of the test signature to the reference signatures of the user i". "High" value of this variable is the fuzzy set B^1 of γ type (see Fig. 8.5), described by the following membership function [42]:

$$\mu_{B^1}(x) = \begin{cases} 0 & \text{for } x \le a \\ \frac{x-a}{b-a} & \text{for } a < x \le b \\ 1 & \text{for } x > b. \end{cases} \tag{8.11}$$

"Low" value is the fuzzy set B^2 of L type (see Fig. 8.5) described by the following membership function [42]:

$$\mu_{B^2}(x) = \begin{cases} 1 & \text{for } x \le a \\ \frac{b-x}{b-a} & \text{for } a < x \le b \\ 0 & \text{for } x > b. \end{cases} \tag{8.12}$$

In the used fuzzy system the value of the parameter a for both rules from the rule base (8.9) is equal to 0 and the value of the parameter b is equal to 1.

- $w_{i,p,r}^{\{s,a\}}$ are weights of the partition associated with the template $\mathbf{tc}_{i,p,r}^{\{s,a\}}$ of the user i, calculated by the formula (8.7).

Verifying the Test Signature

In this method the test signature is recognized as belonging to the user i (genuine) if the assumption $\bar{y}_i > cth_i$ is satisfied, where $\bar{y}_i$ is the value of the output signal of the fuzzy system (4.28):

$$\bar{y}_i \approx \frac{T^*\left\{\begin{array}{c}\mu_{A_{i,1,1}^{1\{v,x\}}}\left(dtst_{i,1,1}^{\{v,x\}}\right),\ldots,\mu_{A_{i,P\{z\},2}^{1\{z,y\}}}\left(dtst_{i,P\{z\},2}^{\{z,y\}}\right);\\ w_{i,1,1}^{\{v,x\}},\ldots,w_{i,P\{z\},2}^{\{z,y\}}\end{array}\right\}}{\left(\begin{array}{c}T^*\left\{\begin{array}{c}\mu_{A_{i,1,1}^{1\{v,x\}}}\left(dtst_{i,1,1}^{\{v,x\}}\right),\ldots,\mu_{A_{i,P\{z\},2}^{1\{z,y\}}}\left(dtst_{i,P\{z\},2}^{\{z,y\}}\right);\\ w_{i,1,1}^{\{v,x\}},\ldots,w_{i,P\{z\},2}^{\{z,y\}}\end{array}\right\}+\\ +T^*\left\{\begin{array}{c}\mu_{A_{i,1,1}^{2\{v,x\}}}\left(dtst_{i,1,1}^{\{v,x\}}\right),\ldots,\mu_{A_{i,P\{z\},2}^{2\{z,y\}}}\left(dtst_{i,P\{z\},2}^{\{z,y\}}\right);\\ w_{i,1,1}^{\{v,x\}},\ldots,w_{i,P\{z\},2}^{\{z,y\}}\end{array}\right\}\end{array}\right)}, \tag{8.13}$$

where $T^*\{\cdot\}$ is the weighted t-norm in the form (4.3) and $cth_i \in [0, 1]$ is a coefficient determined experimentally for each user to eliminate disproportion between FAR and FRR error [43]. The values of this coefficient are usually close to 0.45.

The formula (8.13) results from taking into account in the formula (4.28) (describing a flexible fuzzy system of the Mamdani-type) the following dependencies resulting from the specifics (distribution) of the fuzzy sets shown in Fig. 8.5:

$$\begin{cases} \mu_{B^1}(0) = 0,\ \mu_{B^1}(1) = 1 \\ \mu_{B^2}(0) = 1,\ \mu_{B^2}(1) = 0. \end{cases} \tag{8.14}$$

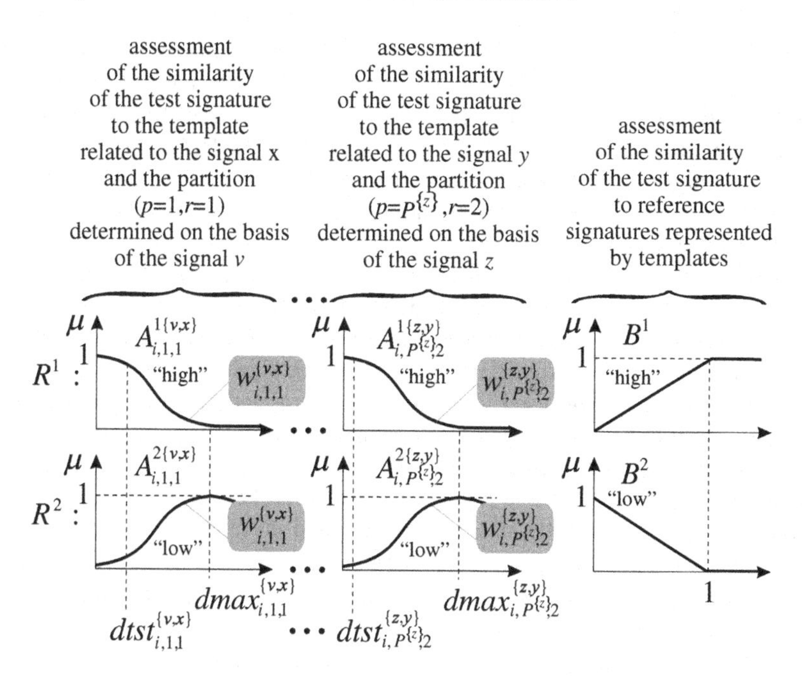

Fig. 8.5 Input and output fuzzy sets used in the rules (8.9) of the flexible fuzzy system for evaluation of similarity of the test signature to the reference signatures

8.1.2 Simulation Results

The comments on the simulations can be summarized as follows:

- The purpose of the simulations was to provide an example of how to improve interpretability of the fuzzy systems aimed at eliminating the learning process in identity verification on the basis of characteristic dynamic signature partitions.
- The information on the used simulation problems is presented in Table 8.1. They are two sample biometric problems.

Table 8.1 A set of test databases used for the simulations on improving fuzzy systems interpretability aimed at eliminating the learning process

Item no.	Name of databases	Number of users	Number of genuine signatures of the user	Number of skilled forged signatures of the user	Problem label
1.	MCYT-100	100	25	25	MCYT
2.	BioSecure	210	15	10	BS

- For each user from the MCYT-100 and BioSecure databases we repeated the training phase and the test phase five times according to the block diagram shown in Fig. 8.1. The results obtained for all users were averaged. In each of the five repetitions we used a different set of training signatures. The described approach is commonly used in evaluating effectiveness of the methods for the dynamic signature verification, which corresponds to the standard crossvalidation procedure (described in Sect. 2.3.2).
- During the training phase we used five randomly selected genuine signatures of each signer. These signatures were used to perform the partitioning, determining of the templates and calculating of the parameters of the fuzzy system for evaluating the similarity of the test signatures to the reference signatures.
- During the test phase we used 15 genuine signatures and 15 forged signatures for the MCYT-100 database. For the BioSecure database we used the 10 remaining genuine signatures and all 10 forged signatures.
- The values of the FAR (False Acceptance Rate) and FRR (False Rejection Rate) errors for the MCYT-100 database are presented in Table 8.2 and those for the BioSecure database are presented in Table 8.4. These errors are used in the literature to evaluate effectiveness of biometric methods. They have been designated for

Table 8.2 Results of the simulations performed by this method using the MCYT-100 database for a different number of sections creating hybrid partitions. The best result is given in bold

$P^{\{s\}}$	$R^{\{s\}}$	Average FAR (%)	Average FRR (%)	Average error (%)
2	**2**	**5.28**	**4.48**	**4.88**
3	2	8.69	5.05	6.87
4	2	8.58	5.73	7.16

Table 8.3 Comparison of the accuracy of different methods for the signature verification for the MCYT-100 database

Method	Average FAR (%)	Average FRR (%)	Average error (%)
Methods of other authors ([11, 13, 14, 20, 29, 32, 33])	-	-	0.74–9.80
Algorithm based on Horizontal Partitioning, AHP ([6])	5.84	5.20	5.52
Algorithm based on Vertical Partitioning, AVP ([5])	5.52	4.87	5.20
Method presented in Sect. 8.1	5.28	4.48	4.88
Method presented in Sect. 8.2	5.72	1.36	3.54

Table 8.4 Results of the simulations performed by this method using the BioSecure database for different number of sections creating hybrid partitions. The best result is given in bold

$P^{\{s\}}$	$R^{\{s\}}$	Average FAR (%)	Average FRR (%)	Average error (%)
2	2	**3.36**	**3.30**	**3.33**
3	2	5.34	5.56	5.45
4	2	9.28	6.16	7.72

Table 8.5 Comparison of the accuracy of different methods for the signature verification for the BioSecure database

Method	Average FAR (%)	Average FRR (%)	Average error (%)
Methods of other authors ([18])	-	-	3.48–30.13
Algorithm based on horizontal partitioning, AHP ([6])	2.94	4.45	3.70
Algorithm based on vertical partitioning, AVP ([5])	3.13	4.15	3.64
Method presented in Sect. 8.1	3.36	3.30	3.33
Method presented in Sect. 8.2	3.29	3.82	3.56

a different number of partitions. Other (less popular) effectiveness measures of biometric methods can also be used (e.g. the ones described in [22]), but in this case it would be difficult to compare the obtained results with the results of other authors.

- The comparison of the accuracy of different methods for the signature verification for the MCYT-100 database is presented in Table 8.3 and for the BioSecure database is presented in Table 8.5.
- The weights of importance of the partitions for the MCYT-100 database are presented in Fig. 8.6 and weights of importance of the partitions for the BioSecure database are presented in Fig. 8.7. Each weight value is the average value of the weights of all users, determined as follows:

$$\bar{w}_{p,r}^{\{s,a\}} = \sum_{i=1}^{I} w_{i,p,r}^{\{s,a\}}. \tag{8.15}$$

The conclusions from the simulations can be summarized as follows:

- For both databases we received the highest accuracy when the signature was divided into two partitions associated with the time moments (initial and final) of signing (see Tables 8.2 and 8.4). By analyzing the results presented in

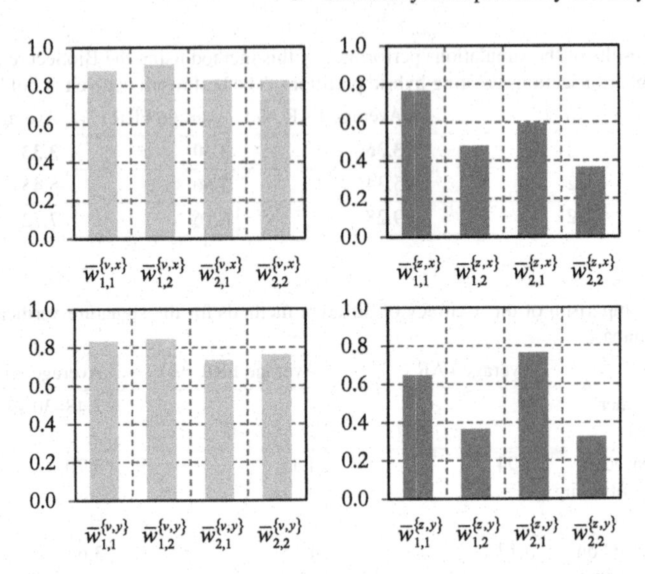

Fig. 8.6 Average values of the partitions weights created for $P^{\{s\}} = 2$ and $R^{\{s\}} = 2$ (variant of partitioning for which the best accuracy was obtained), averaged for all users of the MCYT-100 database

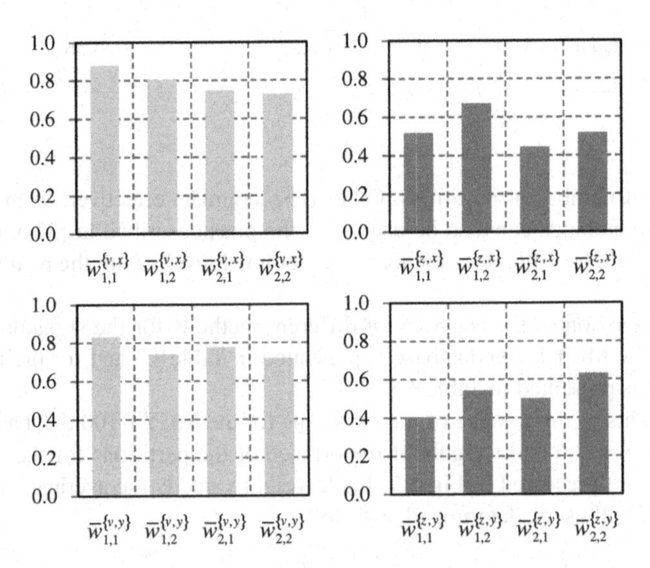

Fig. 8.7 Average values of the partitions weights created for $P^{\{s\}} = 2$ and $R^{\{s\}} = 2$ (variant of partitioning for which the best accuracy has been received), averaged for all users of the BioSecure database

Tables 8.2 and 8.4 it can also be seen that the accuracy of the considered algorithm does not result from increasing the number of sections.

- The algorithm made it possible to select the partitions of the signature in which the reference signatures were created in the most stable way. These partitions had the highest weight value and the lowest value of the parameters describing differences between the reference signatures and the templates in the partitions. It is associated with the lowest tolerance in the evaluation of similarity of the test signatures to the reference signatures. For the MCYT-100 database it was the partition associated with the shape trajectory x, the initial time moment of signing and the high value of velocity (see Fig. 8.6), while for the BioSecure database it was the partition associated with the shape trajectory x, the initial time moment of signing and the low velocity value (see Fig. 8.7). Different results for the two test databases confirm the validity of the algorithm, which adapts its operation to the specificity of the reference signatures, individually for each user.
- For the MCYT-100 database the algorithm promoted the partitions associated with the shape trajectory x, because they had higher weight values (see Fig. 8.6). This shows that for the MCYT-100 database the horizontal movement of the pen was more characteristic during the creation of the reference signatures. In the case of the BioSecure database the algorithm promoted the partitions associated with the shape trajectory x for the signal v and the partitions associated with both shape trajectories (x and y) for the signal z (see Fig. 8.7). This shows that for the BioSecure database the most characteristic of the users are the combinations of (a) the horizontal movement of the pen and the velocity signal value and (b) the vertical movement of the pen and both signals' values.
- For both databases the algorithm promoted the partitions associated with the velocity signal, because they had higher weights values (see Figs. 8.6 and 8.7). It shows that the most characteristic signal describing dynamics of the reference signatures was the velocity signal.
- For both databases the algorithm worked with good accuracy in comparison to other methods for the dynamic signature verification (see Tables 8.3 and 8.5).

8.1.3 Conclusions

In this section we presented the algorithm for the dynamic signature verification using partitioning. The algorithm uses signals available as a standard in graphics tablets. It can be also easily adapted to the specific capabilities of used hardware, e.g. a standard device with a touch screen which is not a graphics tablet. The created partitions are associated with the areas of the signature characterized by: high and low pen velocity and high and low pen pressure on the graphics tablet surface at initial, middle and final moment of the signing process. The algorithm assigns to the partitions weights of importance, which are used in the evaluation of the similarity of the test signatures to the reference signatures. The evaluation is carried out using a fuzzy system in the form (4.28). It uses clear fuzzy rules which allow us to interpret operation of the system. The parameters of the system are determined analytically. It is the most important feature of this algorithm in the process of designing of interpretable fuzzy systems.

8.2 Identity Verification on the Basis of Characteristic Features of the Dynamic Signature

The method discussed in this section represents a so-called global approach to the dynamic signature verification (approaches used in the dynamic signature analysis are presented in Sect. 8.1). This method is based on global features which are extracted from the signature and used during the training and classification phases. It uses the system of the form (4.28) in the verification process of the signature. It is assumed that the structure of the system and its operation are consistent for each user, similar to the system used in Sect. 8.1. Its construction takes into account interpretability criteria described in Sects. 3.4, 4.4 and 6.3. The other characteristic features of the presented method can be summarized as follows:

- It determines individually for each signer weights of importance of the features and takes them into account in the process of signature verification.
- It does not require so-called skilled forgeries and reference signatures of other signers in the training phase (this is a big advantage in this group of methods). It is possible because learning of the fuzzy system used for signature verification is not necessary (it is a one-class classifier like the system presented in Sect. 8.1).
- It is distinguished by the independence of the used set of features which can be arbitrarily reduced or expanded. In other words, the algorithm is flexible because it is not sensitive to the selection of the initial set of features.

8.2.1 Description of the Method

The method contains two (learning and testing) phases, similar to the method outlined in Sect. 8.1. However, actions taken in these phases are different. In the first phase descriptors of features and weights of importance of features are determined. They are needed for the fuzzy system in the test phase to operate properly. These parameters are stored in a database. In the second phase parameters stored for each signer in the learning phase are downloaded from the database. Next, verification of signatures is carried out on the basis of these parameters. In the last part of this section, the learning procedure (Sect. 8.2.1.1) and the signature verification procedure (Sect. 8.2.1.2) are described.

8.2.1.1 Training Phase

The learning phase (**Step 1**) starts by acquiring J reference signatures of the signer i. Different types of tablets may have a different sampling frequency; thus, the acquired signatures should be normalized. The normalization procedure of signatures is shown in Fig. 8.2 and described in Sect. 8.1.

In **Step 2** of the algorithm, the matrix $\mathbf{G}_i$ is determined. The matrix contains values of all global features which describe the dynamics of the reference signatures of the signer i. It has the following structure:

$$\mathbf{G}_i = \begin{bmatrix} g_{i,1,1} & \cdots & g_{i,Nfeat,1} \\ \vdots & & \vdots \\ g_{i,1,J} & \cdots & g_{i,Nfeat,J} \end{bmatrix}, \tag{8.16}$$

where I is the number of signers, J is the number of signatures created by the signer in the acquisition phase, $Nfeat$ is the number of used global features, and $g_{i,n,j}$ is a value of the global feature n, $n = 1, \ldots, Nfeat$, determined for the signature j, $j = 1, \ldots, J$, created by the signer i, $i = 1, \ldots, I$. The method of determining values of the global features is described in detail in [14] and it will not be considered in this section.

In **Step 3** of the algorithm, the vector $\bar{\mathbf{g}}_i = \left[\bar{g}_{i,1}, \ldots, \bar{g}_{i,Nfeat}\right]$ is determined, where $\bar{g}_{i,n}$ is an average value of n-th global feature of all J reference signatures of the signer i:

$$\bar{g}_{i,n} = \frac{1}{J} \cdot \sum_{j=1}^{J} g_{i,n,j}. \tag{8.17}$$

In **Step 4** of the algorithm, calculation of the fuzzy system parameters used in the test phase is performed. This procedure is called `Classifier Determination` $(i, \mathbf{G}_i, \bar{\mathbf{g}}_i)$. In particular, distances $maxd_{i,n}$ and weights $w_{i,n}$ $(i = 1, \ldots, I,\ n = 1, \ldots, Nfeat)$ are determined individually for the signer i. Each parameter $maxd_{i,n}$ determines instability of signing of the signer i in the context of the feature n. Its value is dependent on the variability of the feature. Each weight $w_{i,n}$ describes importance of the global feature n.

In **Step 5** of the algorithm, the following information about the signer i is stored into a database: the vector $\bar{\mathbf{g}}_i$ and parameters of the fuzzy system $maxd_{i,n}$ and $w_{i,n}$.

The key step in the described algorithm is the procedure of `Classifier Determination` $(i, \mathbf{G}_i, \bar{\mathbf{g}}_i)$. It is executed in Step 4, which is used for determining fuzzy system parameters and it uses matrix $\mathbf{G}_i$ and vector $\bar{\mathbf{g}}_i$. This procedure starts by determining Euclidean distances $d_{i,n,j}$ between each global feature n and the average value of the global feature for all J signatures of the signer i (**Step 4.1**):

$$d_{i,n,j} = \left|\bar{g}_{i,n} - g_{i,n,j}\right|. \tag{8.18}$$

In **Step 4.2** of the considered procedure, selection of the maximum distance for each global feature n is performed (from the distances determined in Step 4.1):

$$dmax_{i,n} = \max_{j=1,\dots,J} \left\{ d_{i,n,j} \right\}. \tag{8.19}$$

If reference signatures are more similar to each other, the tolerance of the used fuzzy system (4.28) is lower, because $dmax_{i,n}$ takes smaller values. In **Step 4.3** of the considered procedure, computation of weights $w_{i,n}$ is performed. Each weight is calculated on the basis of the standard deviation of n-th global feature of the signer i and the average value of distances for n-th feature of the signer i:

$$w_{i,n} = 1 - \frac{\sqrt{\frac{1}{J} \cdot \sum_{j=1}^{J} {d_{i,n,j}}^2}}{\frac{1}{J} \cdot \sum_{j=1}^{J} d_{i,n,j}}. \tag{8.20}$$

It should be emphasized that the distances and the weights are used in the signature verification phase.

8.2.1.2 Test Phase (Verification of Signatures)

The purpose of the signature verification phase is to determine whether the tested signature which belongs to a signer claiming to be the signer i in fact belongs to the signer i. The verification phase is performed in a similar way as in Sect. 8.1.1.2.

In **Step 1** of the procedure a signer, whose identity should be verified, creates one test signature. In this step he/she also claims his/her identity as i. As in the case of the learning phase, the signature has to be geometrically pre-processed.

In **Step 2** of the procedure, the following information is downloaded from the database: average values of global features calculated during the training phase ($\bar{\mathbf{g}}_i$) and fuzzy system parameters of the signer i ($dmax_{i,n}$, $w_{i,n}$).

In **Step 3** of the procedure, determination of the values of the global features $gtst_{i,n}$, $n = 1, \dots, Nfeat$, for the test signature is performed.

In **Step 4** of the procedure, similarities of global features values of the test signature to the average values of the global features for the reference signatures are determined:

$$dtst_{i,n} = \left| \bar{g}_{i,n} - gtst_{i,n} \right|. \tag{8.21}$$

In the last step (**Step 5**) of the procedure the verification of the test signature using fuzzy system of the form (4.28) (described in Sect. 4.3) is performed. The values of the signals $dtst_{i,n}$ determined in Step 4 are given at the input of the fuzzy system.

Table 8.6 A set of symbols used in the methods discussed in Sects. 8.1 and 8.2

Item no.	Interpretation of parameter	Symbol of parameter in Sect. 8.1	Symbol of parameter in Sect. 8.2
1.	Border of inclusion of features (similarity of shape signal, global feature values, etc.) of reference signatures	$dmax_{i,p,r}^{\{s,a\}}$	$dmax_{i,n}$
2.	Feature value (similarity of shape signal, global feature values, etc.) of test signature	$dtst_{i,p,r}^{\{s,a\}}$	$dtst_{i,n}$
3.	Weight of importance of a feature (similarity of shape signal, global feature values, etc.) determined for reference signatures	$w_{i,p,r}^{\{s,a\}}$	$w_{i,n}$

Operation of the algorithms discussed in Sects. 8.1 and 8.2 has some similarities. They are detailed in Table 8.6. Taking into account these analogies we resigned from a detailed description of the verification phase in this section. In applications of fuzzy systems in which the learning process can be eliminated counterparts of the variables listed in Table 8.6 will usually need to be indicated.

8.2.2 *Simulation Results*

Comments on the simulations can be summarized as follows:

- The purpose of the simulations was to show an example of improving interpretability of fuzzy systems aimed at eliminating the learning process in identity verification on the basis of characteristic dynamic signature's global features.
- The information on the used simulation problems is presented in Table 8.1. They are the same problems which were discussed in Sect. 8.1.

The conclusions from the simulations can be summarized as follows:

- The method for the considered databases works with high accuracy in comparison with the methods presented in Tables 8.3 and 8.5. The comparison criterion was the value of the EER (Equal Error Rate) error, which is commonly used to evaluate accuracy of biometric methods [11, 25].

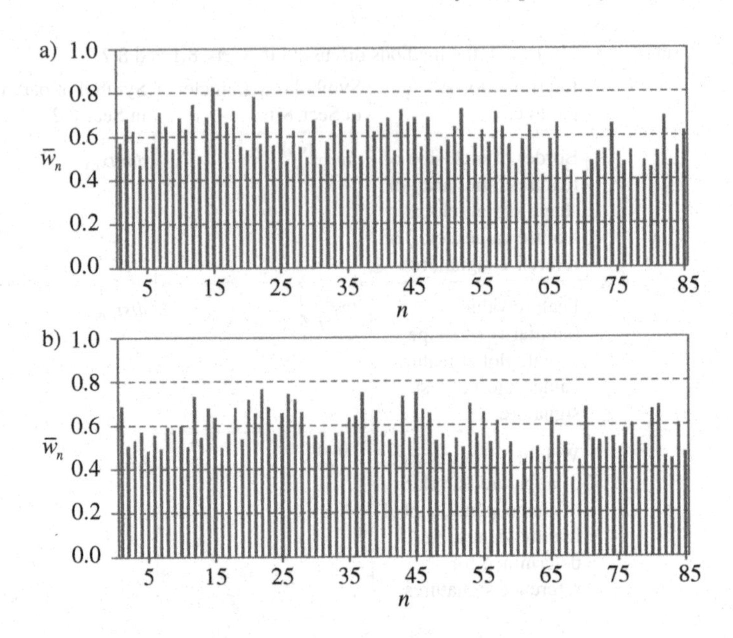

Fig. 8.8 Average values of weights of the global for: **a** MCYT, **b** BS

- In the simulations a common value of $cth_i = 0.45$ was used for all signers. If the algorithm operating in practice has to be, e.g., more sensitive to false acceptance (e.g. in high security systems), the value of cth_i should be higher than 0.45.

8.2.3 Conclusions

In this section we have discussed a method for the dynamic signature verification using global features. It determines weights of importance of the global features and uses them in the signature verification process (Fig. 8.8). It is also worth noting that the considered algorithm works independently of the initial set of features, works without access to the so-called skilled forgeries and uses the capabilities of the fuzzy system whose knowledge can be interpreted. The discussed fuzzy system, like the system presented in Sect. 8.1, does not require learning.

8.3 Summary

In this chapter we have considered ways of assuring interpretability of fuzzy systems by elimination of the supervised learning. They result from basic assumptions on fuzzy systems, under which the rules of fuzzy systems can be formulated by experts.

However, the main assumptions of the solutions discussed in this chapter concern initialization of a determined form of fuzzy rules for the given problem and next, an analytical determination of their parameters on the basis of learning data. An advantage of this approach is a possibility of interpretation of each parameter in the fuzzy rules base.

The algorithms considered in this chapter seem to be working in accordance with a certain scheme, which can be duplicated in other application areas provided we have a set of training data. Those algorithms have been proposed in the papers [7, 46]. It should be emphasized that the next step towards increasing the interpretability of fuzzy systems used in the algorithms can be e.g. reduction of the number of input attributes of fuzzy system (i.e. the number of partitions of the dynamic signature or the number of global features of the dynamic signature). This reduction can take into account e.g. a simple selection on the basis of weights values or an automatic selection implemented using, for example, a population-based algorithm [45], the PCA algorithm [47], etc. The pre-obtained results seem to be interesting.

As already mentioned, some aspects of assuring interpretability of fuzzy systems designed with elimination of the supervised learning are partially considered in our previous papers (see e.g. [5–7]).

References

1. Bankó, Z., János, A.: Correlation based dynamic time warping of multivariate time series. Expert Syst. Appl. **39**, 12814–12823 (2012)
2. Batista, L., Granger, E., Sabourin, R.: Dynamic selection of generative discriminative ensembles for off-line signature verification. Pattern Recognit. **45**, 1326–1340 (2012)
3. Beltrán, M., Melin, P., Trujillo, L.: Modular neural networks with fuzzy response integration for signature recognition. Bio-inspired Hybrid Intelligent Systems for Image Analysis and Pattern Recognition, vol. 256, pp. 81–91. Springer, Heidelberg (2009)
4. Carrera, V., Melin, P., Bravo, D.: Development of an automatic method for classification of signatures in a recognition system based on modular neural networks. Recent Advances on Hybrid Intelligent Systems, vol. 451, pp. 201–210. Springer, Heidelberg (2013)
5. Cpałka, K., Zalasiński, M.: On-line signature verification using vertical signature partitioning. Expert Syst. Appl. **41**, 4170–4180 (2014)
6. Cpałka, K., Zalasiński, M., Rutkowski, L.: New method for the on-line signature verification based on horizontal partitioning. Pattern Recognit. **47**, 2652–2661 (2014)
7. Cpałka, K., Zalasiński, M., Rutkowski, L.: A new algorithm for identity verification based on the analysis of a handwritten dynamic signature. Appl. Soft Comput. **43**, 47–56 (2016)
8. Dean, D., Sridharan, S.: Dynamic visual features for audio-visual speaker verification. Comput. Speech Lang. **24**, 136–149 (2010)
9. De Fernandez, J., Su, C.J., Chiang, C.Y., Huang, J.Y.: Kinect-enabled home-based rehabilitation system using dynamic time warping and fuzzy logic. Appl. Soft Comput. **22**, 652–666 (2014)
10. Ekinci, M., Aykut, M.: Human gait recognition based on kernel PCA using projections. J. Comput. Sci. Technol. **22**, 867–876 (2007)
11. Faundez-Zanuy, M.: On-line signature recognition based on VQ-DTW. Pattern Recognit. **40**, 981–992 (2007)
12. Faundez-Zanuy, M., Pascual-Gaspar, J.M.: Efficient on-line signature recognition based on multi-section vector quantization. Form. Pattern Anal. Appl. **14**, 37–45 (2011)

13. Fierrez, J., Ortega-Garcia, J., Ramos, D., Gonzalez-Rodriguez, J.: HMM-based on-line signature verification: feature extraction and signature modeling. Pattern Recognit. Lett. **28**, 2325–2334 (2007)
14. Fierrez-Aguilar, J., Nanni, L., Lopez-Penalba, J., Ortega-Garcia, J., Maltoni, D.: An on-line signature verification system based on fusion of local and global information. Lecture Notes in Computer Science, vol. 3546, pp. 523–532. Springer, Heidelberg (2005)
15. Galbally, J., Diaz-Cabrera, M., Ferrer, M.A., Gomez-Barrero, M., Morales, A., Fierrez, J.: On-line signature recognition through the combination of real dynamic data and synthetically generated static data. Pattern Recognit. **48**, 2921–2934 (2015)
16. Gonzalez, R.C., Woods, R.E.: Digital Image Processing. Pearson Education Inc., Singapore (2002)
17. Guerbai, Y., Chibani, Y., Hadjadji, B.: The effective use of the one-class SVM classifier for handwritten signature verification based on writer-independent parameters. Pattern Recognit. **48**, 103–113 (2015)
18. Houmani, N., Garcia-Salicetti, S., Mayoue, A., Dorizzi, B.: BioSecure Signature Evaluation Campaign 2009 (BSEC'2009). http://biometrics.it--sudparis.eu/BSEC2009/downloads/BSEC2009_results.pdf Accessed 15 Oct 2015
19. Huang, K., Hong, Y.: Stability and style-variation modeling for on-line signature verification. Pattern Recognit. **36**, 2253–2270 (2003)
20. Ibrahim, M.T., Khan, M.A., Alimgeer, K.S., Khan, M.K., Taj, I.A., Guan, L.: Velocity and pressure-based partitions of horizontal and vertical trajectories for on-line signature verification. Pattern Recognit. **43**, 2817–2832 (2010)
21. Jain, A.K., Griess, F.D., Connell, S.D.: On-line signature verification. Pattern Recognit. **35**, 2963–2972 (2002)
22. Jain, A.K., Ross, A.: Introduction to biometrics. In: Jain, A.K., Flynn, P., Ross, A.A. (eds.) Handbook of Biometrics, pp. 1–22. Springer, US (2008)
23. Jeong, Y.S., Jeong, M.K., Omitaomu, O.A.: Weighted dynamic time warping for time series classification. Pattern Recognit. **44**, 2231–2240 (2011)
24. Khan, M.A.U., Khan, M.K., Khan, M.A.: Velocity-image model for online signature verification. IEEE Trans. Image Process. **15**, 3540–3549 (2006)
25. Kholmatov, A., Yanikoglu, B.: Identity authentication using improved online signature verification method. Pattern Recognit. Lett. **26**, 2400–2408 (2005)
26. Kumar, R., Sharma, J.D., Chanda, B.: Writer-independent off-line signature verification using surroundedness feature. Pattern Recognit. Lett. **33**, 301–308 (2012)
27. Lee, C.P., Leu, Y., Yang, W.N.: Constructing gene regulatory networks from microarray data using GA/PSO with DTW. Appl. Soft Comput. **12**, 1115–1124 (2012)
28. Lei, H., Govindaraju, V.: A comparative study on the consistency of features in on-line signature verification. Pattern Recognit. Lett. **26**, 2483–2489 (2005)
29. Lumini, A., Nanni, L.: Ensemble of on-line signature matchers based on overcomplete feature generation. Expert Syst. Appl. **36**, 5291–5296 (2009)
30. Maiorana, E.: Biometric cryptosystem using function based on-line signature recognition. Expert Syst. Appl. **37**, 3454–3461 (2010)
31. Moon, J.H., Lee, S.G., Cho, S.Y., Kim, Y.S.: A hybrid online signature verification system supporting multi-confidential levels defined by data mining techniques. Int. J. Intell. Syst. Technol. Appl. **9**, 262–273 (2010)
32. Nanni, L.: An advanced multi-matcher method for on-line signature verification featuring global features and tokenised random numbers. Neurocomputing **69**, 2402–2406 (2006)
33. Nanni, L., Lumini, A.: Ensemble of Parzen window classifiers for on-line signature verification. Neurocomputing **68**, 217–224 (2005)
34. Nanni, L., Lumini, A.: Advanced methods for two-class problem formulation for on-line signature verification. Neurocomputing **69**, 854–857 (2006)
35. Nanni, L., Lumini, A.: A novel local on-line signature verification system. Pattern Recognit. Lett. **29**, 559–568 (2008)

36. Nanni, L., Maiorana, E., Lumini, A., Campisi, P.: Combining local, regional and global matchers for a template protected on-line signature verification system. Expert Syst. Appl. **37**, 3676–3684 (2010)
37. O'Reilly, Ch., Plamondon, R.: Development of a Sigma-Lognormal representation for on-line signatures. Pattern Recognit. **42**, 3324–3337 (2009)
38. Parodi, M., Gómez, J.C.: Legendre polynomials based feature extraction for online signature verification. Consistency analysis of feature combinations. Pattern Recognit. **47**, 128–140 (2014)
39. Pascual-Gaspar, J.M., Faúndez-Zanuy, M., Vivaracho, C.: Fast on-line signature recognition based on VQ with time modelling. Eng. Appl. Artif. Intell. **24**, 368–377 (2011)
40. Radhika, K.S., Gopika, S.: Online and offline signature verification: a combined approach. Procedia Comput. Sci. **46**, 1593–1600 (2015)
41. Radhika, K.R., Venkatesha, M.K., Sekhar, G.N.: Signature authentication based on subpattern analysis. Appl. Soft Comput. **11**, 3218–3228 (2011)
42. Rutkowski, L.: Computational Intelligence. Springer, Heidelberg (2010)
43. Yeung, D.Y., Chang, H., Xiong, Y., George, S., Kashi, R., Matsumoto, T., Rigoll, G.: SVC2004: First international signature verification competition. Lecture Notes in Computer Science, vol. 3072, pp. 16–22. Springer, Heidelberg (2004)
44. Zalasiński, M., Cpałka, K.: New approach for the on-line signature verification based on method of horizontal partitioning. Lecture Notes In Artificial Intelligence, vol. 7895, pp. 342–350. Springer, Heidelberg (2013)
45. Zalasiński, M., Cpałka, K., Hayashi, Y.: New method for dynamic signature verification based on global features. Lecture Notes in Computer Science, vol. 8467, pp. 251–265. Springer, Switzerland (2014)
46. Zalasiński, M., Cpałka, K., Hayashi, Y.: New fast algorithm for the dynamic signature verification using global features values. Lecture Notes in Computer Science, vol. 9120, pp. 175–188. Springer, Switzerland (2015)
47. Zalasiński, M., Łapa, K., Cpałka, K.: New algorithm for evolutionary selection of the dynamic signature global features. Lecture Notes In Artificial Intelligence, vol. 7895, pp. 113–121. Springer, Heidelberg (2013)

Chapter 9
Concluding Remarks and Future Perspectives

In this book we have presented solutions for designing interpretable fuzzy systems. Chapter 1 outlines the subject of the book and the contents of the individual chapters. Chapters 2 and 3 are introductory chapters. The first one, i.e. Chap. 2, describes a number of selected issues in the field of fuzzy systems. We place a particular focus on those issues which contribute to proper understanding of the issues presented in the subsequent chapters. In Chap. 3 we review issues related to interpretability of fuzzy systems. This chapter identifies and analyzes the current trends connected with improving interpretability. In Chap. 4 we consider possibilities of affecting interpretability of fuzzy systems by appropriate selection of their structure. This is important because an appropriate selection of the structure of a fuzzy system makes it possible to increase its precision, and thus, simplify the fuzzy rule base. In Chap. 5 we describe possibilities of improving interpretability of fuzzy systems designed using the gradient learning. In this context these systems allow us to combine high-precision of gradient algorithms with high efficiency of initialization, reduction, merging and tuning algorithms. In Chap. 6 we discuss possibilities of improving interpretability of fuzzy systems designed using evolutionary learning. Here, these systems make it possible to combine high flexibility of population-based algorithms with often rather extensive expectations concerning interpretability, which can be expressed in the form of appropriate criteria. In Chap. 7 we discuss how to improve interpretability of fuzzy systems applied to two practical problems: (a) problem of weakly nonlinear dynamic objects modelling and (b) the problem of control of a CNC machine with a jerk limit. Proposed improvements provide an opportunity for an unconventional use of fuzzy systems, which involves, among others, integration of fuzzy systems' capabilities with some other solutions which prove to work well in practice and do not use the fuzzy sets theory. The develop method requires an individual approach; however, it can result in a precise use of a fuzzy system, and thus simplification of its fuzzy rule base. In Chap. 8 we describe possibilities of improving interpretability of fuzzy systems by eliminating the supervised learning. These possibilities show an alternative way of using learning (input) data, if available. The main objective of this

K. Cpałka, *Design of Interpretable Fuzzy Systems*, Studies in Computational Intelligence 684, DOI 10.1007/978-3-319-52881-6_9

approach is to use data in such way so as not to lose the possibility of interpreting any parameter of the system which determines the form of fuzzy rules.

Therefore, the purpose of this book has been to present the important issue of fuzzy systems interpretability from different points of view. It is an extensive problem and it cannot be reduced only to the issue of readability of fuzzy sets and rules. For this reason, the topics covered in this book do not exhaust the subject of our considerations, because there are still problems which are waiting for effective solutions to be found and their examples include the following problems:

- Construction of dedicated algorithms for interpretable fuzzy systems designing. They can e.g. extract a set of non-dominated solutions, take into account new exploration and exploitation operators, use new rules of cooperation between populations. They can also provide a better adjustment of their operation to the specifics of a rule-based notation and flexibility aspects. In paper [5] we propose a sample population-based algorithm which was created by combining the genetic algorithm and genetic programming. Moreover, a systematic arrangement of solutions from the field of population-based algorithms which have been presented in the literature seems to be an interesting challenge. They could be important tools for improving interpretability.
- Construction of interpretable fuzzy systems for online applications. These systems can be used, e.g., for classification of stream data incoming in the real time, whose theoretical length can be infinite. They can also be used for modelling processes and objects which are the source of data obtained from identification. In construction of these systems we can take into account a variety of factors including flexibility aspects and solutions aimed at eliminating the learning process. In paper [4] we propose a preliminary sample approach for automatic construction of fuzzy systems for online applications. It takes into account an analysis of fuzzy rules activity.
- Construction of solutions used for design of hierarchical type interpretable fuzzy systems and fuzzy systems processing imprecise information. Construction of these systems can also involve various interpretability aspects, which are considered in the previous chapters of the book. This approach could certainly contribute to the development of many interesting scientific solutions of high practical importance.
- Construction of different criteria taken into account in interpretable fuzzy system designing. In the previous chapters of this book we focus on interpretability criteria of fuzzy systems. New criteria of this type can be created, but it needs to be noted that also new criteria can be created which enforce e.g. better adjustment of fuzzy systems to their hardware implementation (e.g. in the form of emulators of objects or processes). In papers [6, 7] we initially considered sample solutions developed in the context of FPGA technology (Field-Programmable Gate Array).
- Use of the potential of interpretable fuzzy systems in new applications. It seems that hybrid solutions in which possibilities of fuzzy systems are complemented by possibilities of other proven methods have a special application importance.

Examples of these solutions are presented in Chaps. 7 and 8 as well as in our previous papers [1, 8, 9].

- Use of solutions in a wide variety of issues from the field of interpretable fuzzy systems designing in other areas of applications. They are issues in which we first need to indicate the structure of a solution, taking into account certain criteria, and then determine parameters of the structure. They are solved the most often by the trial and error method (this especially applies to the choice of structure), and therefore, the hybrid solutions proposed in the previous chapters are of high practical importance. In our previous papers we provide examples illustrating how to use them when designing regulators implementing the PID algorithm (Proportional-Integral-Derivative) [2, 3]. However, we can design other complex systems in the same way including filter systems, neural networks structures, etc. From a practical point of view attempting to automate such design processes could certainly be an interesting and important challenge.

References

1. Bartczuk, Ł., Przybył, A., Cpałka, K.: A new approach to nonlinear modelling of dynamic objects based on fuzzy rules. Int. J. Appl. Math. Comput. Sci. **26**, 603–621 (2016)
2. Cpałka, K., Łapa, K., Przybył, A.: A new approach to design of control systems using genetic programming. Inf. Technol. Control **44**, 433–442 (2015)
3. Łapa, K., Cpałka, K.: On the application of a hybrid genetic-firework algorithm for controllers structure and parameters selection. Advances Intelligent Systems and Computing, pp. 111–123. Springer, Switzerland (2016)
4. Łapa, K., Cpałka, K., Hayashi, Y.: New approach for nonlinear modelling based on online designing of the fuzzy rule base. Lecture Notes in Computer Science, vol. 9692, pp. 230–247. Springer, Switzerland (2016)
5. Łapa, K., Cpałka, K., Koprinkova-Hristova, P.: New method for fuzzy nonlinear modelling based on genetic programming. Lecture Notes in Computer Science, vol. 9692, pp. 432–449. Springer, Switzerland (2016)
6. Przybył, A., Er, M.J.: A new approach to designing of intelligent emulators working in a distributed environment. Lecture Notes in Computer Science, vol. 9693, pp. 546–558. Springer, Switzerland (2016)
7. Przybył, A., Er, M.J.: The method of hardware implementation of fuzzy systems on FPGA. Lecture Notes in Computer Science, vol. 9692, pp. 284–298. Springer, Switzerland (2016)
8. Rutkowski, L., Przybył, A., Cpałka, K.: Novel on-line speed profile generation for industrial machine tool based on flexible neuro-fuzzy approximation. IEEE Trans. Ind. Electron. **59**, 1238–1247 (2012)
9. Zalasiński, M., Cpałka, K., Rakus-Andersson, E.: An idea of the dynamic signature verification based on a hybrid approach. Lecture Notes in Computer Science, vol. 9693, pp. 232–246. Springer, Switzerland (2016)

Index

K. Cpałka, *Design of Interpretable Fuzzy Systems*, Studies in Computational Intelligence 684, DOI 10.1007/978-3-319-52881-6

Printed in the United States
By Bookmasters